Computational
Methods for
Data Analysis

A WILEY PUBLICATION IN APPLIED STATISTICS

Computational Methods for Data Analysis

JOHN M. CHAMBERS
Bell Laboratories
Murray Hill, New Jersey

JOHN WILEY & SONS, New York • Chichester • Brisbane • Toronto

Published by John Wiley & Sons, Inc.

Library of Congress Cataloging in Publication Data

Chambers, John M
Computational methods for data analysis.

(Wiley series in probability and mathematical statistics)
Includes index.
1. Mathematical statistics—Data processing.
2. Numerical analysis—Data processing. I. Title.

QA276.4.C48 519.4 77-9493
ISBN 0-471-02772-3

Printed in the United States of America

10 9 8 7 6 5 4 3 2

for Bea

PREFACE

The material in this book covers the major computational methods which are important for data analysis. The approach is to present the essential results on each topic, including an appraisal of currently competitive methods and references to selected algorithms. A special Appendix gathers together these references, with information on the language and format of the algorithms and how to obtain them. Most chapters conclude with a set of problems. These are intended to be of practical value to the reader. None is entirely trivial and some involve designing a significant set of algorithms.

Readers looking for a specific topic should go first to the quick reference on the inside back cover. This will give page references for both the method and the algorithms. If the topic of interest is not obviously covered in the quick reference, try the index.

This is the first book to attempt a complete picture of computing for data analysis. I hope the result will be useful for several groups of people. Those involved in analysing data may look here to find computing methods currently competitive for specific problems. Users of existing packages and programs may compare the reliability, accuracy and cost of the methods used to the current best (and, I hope, will react strongly to any defects they find in the packages).

Those who are or would like to be active in statistical computing will find introductions to various topics, a wealth of unsolved or partially solved problems, and references to more detailed treatments. (Some of the important areas for future work are listed in the Index, under "open problems".) Finally, I hope professionals in the various fields of computing science will find it useful to see the viewpoint and needs suggested by data analysis.

Computing for data analysis is an important, challenging and varied field. Data analysis is increasingly an important part of scientific research, economic analysis and many other activities. Major analyses usually require extensive computing; hence, the validity and adequacy of the computation must be demonstrable to support the analysis. An awareness of the best computing methods and of their limitations will be increasingly important. It is essential, when evaluating computing techniques, to be aware as well of the real goals of the data analysis, so as to sort out the relevant from the merely interesting. Having invoked these important responsibilities, I should quickly add that working on these frontiers of computing and data analysis is frequently exhilarating and great fun.

I have had the opportunity of working at one time or another in all the major areas described in this book. As a result, any list of indebtedness would be sure to offend by omission. I can only hope that the many experts who have given generously of their time and advice will detect some beneficial effects in the result. The support of Bell Laboratories in the research and writing of this book is gratefully acknowledged. I am particularly grateful for the stimulation of discussions with many colleagues there, past and present. The entire contents of the book have been set on a computer photocomposition system at Bell Laboratories. It is hoped that this support will both add to the timeliness of the material and allow some cost savings to be passed on to buyers. Portions of the material have been presented in courses at Harvard University, Princeton University and Bell Laboratories. Comments and questions from students have been valuable in subsequent revisions.

Note to the Reader

The Chapters of the book are as self-contained as reasonably possible. References to Equations, Figures and Sections within the Chapter are given without Chapter number. The References are collected, by Chapter, beginning on page 228. The Appendix deals with algorithms by Chapter.

CONTENTS

Computational

Methods for

Data Analysis

CHAPTER ONE

Introduction

a. General Approach

This book is intended to assist those who are involved in the planning, selection, or development of computational support for data analysis. There is general recognition of the importance of such support for effective research and development in all branches of science. The ability to do massive amounts of programmed computations is the foremost change in the intellectual environment in the second half of the twentieth century. The effective amount of calculation possible in a given time has increased by about five orders of magnitude. Several revolutionary changes in equipment and, to a somewhat lesser extent, in programming facilities have entirely altered the scope of computing. The capabilities and practical economics of computing continue to change at a rapid rate in many respects.

Chapters 2 through 8 treat major topics of computational technique. Chapter 2 discusses programming and program evaluation in general. Chapter 3 deals with data management and related problems. Chapters 4 through 7 cover numerical methods (general methods, linear methods, nonlinear fitting, and simulation). Chapter 8 discusses graphical computation.

A wide variety of computational topics is treated, since data analysis needs support from nearly every branch of computer science. The simpler approach of concentrating on a few, relatively straightforward topics will not provide sufficient guidance for practical work. The computational topics are treated from the special viewpoint of data analysis. The emphasis, techniques, and recommendations, therefore, differ significantly from those for other applications.

For each topic, the discussion identifies the essential computational problems, outlines and compares the currently competitive approaches and provides references to more detailed material, including reliable published or generally available algorithms where possible. For the most part, it is assumed that some programming effort will be required, either in implementing procedures or in modifying existing algorithms or other special-purpose software.

No attempt is made to rate existing packages or languages for statistical computing in any overall sense. Users will be able to check the facilities provided against the currently competitive methods for a particular problem. Also, the steps required to extend or improve the current capabilities of the package can be assessed from the descriptions here. The concern is to outline what is currently known and available to solve specific problems. Users of packages may assess the facilities provided against this background. We especially try to suggest how new or extended facilities can be developed.

Both current abilities and long-range prospects vary greatly among different areas of computing. Some problems are relatively easy and can be considered to be understood, in broad outline at least; for example, linear least-squares and sorting. Others are in a less satisfactory state at the moment, but can in principle be well handled: for example, provision of language interfaces and graphical computations. Finally, some areas are intrinsically too complex ever to have single, general solutions. For such areas, a vague general problem must be sharpened to make it computationally meaningful. Examples are the simulation of random numbers and, possibly, general fitting of nonlinear models.

b. Data Analysis and Computing

As a general preliminary to detailed discussions, it is useful to characterize possible computing environments and the kinds of data analysis likely to arise. Such questions affect the feasibility or relevance of techniques to be discussed.

The data analysis may be characterized by such questions as:

(1) Is it primarily routine (consisting of specialized and repetitive analyses) or primarily exploratory (unpredictable as to the kind of problem and the appropriate response)?

(2) Are the problems large or small, both in terms of the volume of data and of the resources (personnel, computing facilities, etc.) available?

Exploratory analysis places greater emphasis on the flexibility and extensibility of the computing facilities. Routine analysis allows greater emphasis on efficiency and ease of standard use of programs, and permits greater effort to be made in refining computing procedures. Generally, routine analysis favors specialized systems, while exploratory analysis favors general, open-ended support.

Small projects will not be as likely to undertake major new programming efforts. The primary emphasis will be on quick and easy access to reasonably appropriate computing tools. At the same time, direct checking of results and adhoc revisions and interpretation will be more feasible on small bodies of data. Larger projects may not be justified in compromising the details of their analysis or their computations, while large data sets always require great care in checking for errors and misinterpretations. Greater control over the details of computation will then be required.

A different characteristic of the environment is the general style in which computing for data analysis is presented. In Section **2.b** some criteria for evaluating individual algorithms and systems are discussed. A more general characteristic is the contrast between special-purpose statistical systems and general-purpose, nonstatistical languages. The merits of these have been the focus of considerable debate. The two extremes are a statistical system that replaces all programming in general languages and direct programming in a general language. Intermediate forms include statistical systems that permit the kind of programming done in ordinary languages, and programming in general languages with extensive support from subroutine libraries and other facilities. Special-purpose systems can provide commonly used analyses or data structures as simple primitives; initial and routine use is facilitated. General-purpose approaches have advantages of extensibility and flexibility. In particular, the incorporation of new procedures written elsewhere is usually more straightforward.

In assessing various approaches to computing for data analysis, it is relevant to consider changes in the form of user communication with computers and changes in the population of potential users since initial development of statistical systems. In the typical computer facility until 1965, user communication was by an externally prepared program, usually on punched cards, presented for a relatively slow run (with several hours or more elapsing before obtaining results). Such an environment discourages innovative, flexible use of computer systems. Subsequent developments have led to computer access via interactive terminals with essential support of text editors and other non-numerical systems. Statistical computations also typically are executed more rapidly as well, either in an interactive system, or by a spawned job with a short time to completion. The new style encourages direct and innovative participation by data analysts in computing. A contemporary development is the training of a generation of data analysts who had early exposure to computing and who accept computer programming as a natural expression of the analysis they intend to perform. These developments are relevant to planning because better user communication and data analysts who are at ease with programming allow, and often demand, broader and more flexible computer support. Some restrictive package systems, for example, were suitable for the earlier environment but needed to be revised to be acceptable subsequently.

For future planning, however, the best features of both approaches should be sought. Building upon a widely available general programming language (probably FORTRAN), and taking advantage of high-quality existing algorithms, will provide a strong problem-solving support. Convenient user interfaces, where needed, may be built upon this basis, using modern techniques of language design. (See Sections **2.d** to **2.f**.)

CHAPTER TWO

Programming

This chapter considers some questions about programs that apply across most areas of technique. What good features should programs have? How can competing algorithms be compared? What principles underlie the design and writing of good programs? What features of programming languages are relevant to computing for data analysis? In the following sections, a few points are discussed that relate to these questions.

a. Structure of Computations in Data Analysis

The process of data analysis, whether or not it is carried out by computer, has a basic structure that closely parallels the operation of a computer program. The sequence of steps in each case involves the same four components: the acquisition of external data (input), the planning and definition of the analysis (programming), the execution of the analysis (calculation), and the display of results and summaries (output).

These components define the areas that must be considered when developing a computational capability for data analysis. At the same time, they are each areas in which much development of computing technology and methods has occurred. The provision of effective data analysis requires some familiarity with recent developments in these areas. Later chapters present the important current techniques in various specialized areas of computing. The present chapter attempts to take a general look at the process of providing computer programs.

General advice on the planning and implementation of statistical computing facilities should be given with modesty and caution. Each organization differs in the goals and constraints applied to its data analysis, in the relative priorities it assigns to these, and in the time-scale on which these goals and constraints are to be considered.

At the same time, the resources available will also differ, in terms of the number and skills of personnel, the quantity and scope of computing hardware, and the extent and usefulness of existing programs.

Also, the computational requirements for data analysis differ from, but still overlap with, the needs of other computer applications. Therefore, an approach that treats data analysis as if it were the sole user of a computer system may overlook the benefits of planned cooperation with other groups of computer users.

The approach to computers will certainly depend on all these local conditions. However, there are useful general concepts to be applied in assessing each particular case. By keeping such concepts in mind, planners may achieve a more relevant and adequate facility within their environment.

The first principle to establish is that the computational needs are part of the whole process of data analysis. The effectiveness of a particular computing tool, therefore, lies in its overall role in the analysis to which it is applied. The costs and benefits must be balanced in this broader context, not just in terms of the computing itself. This principle, apparently self-evident, is quite difficult to apply consistently in practice, as is shown by many of the examples cited in later chapters.

The next questions to answer are then: "What good things can computers provide in data analysis?" and "What difficulties and costs can be associated with computing?" It is possible to give some general, although incomplete, answers to these questions.

The *benefits* of good computing support for data analysis are of great magnitude. While most of them (increased speed of computation, ability to handle larger data bases, the availability of new and better algorithms) may be regarded as quantitative, they are often sufficiently major to revolutionize the entire approach to data analysis. In many applications, however, the full benefit of modern computing methods will be obtained only when the traditional statistical methodology and concepts are rethought. Frequently, the result is a more natural and unrestricted attitude to the methods themselves, uninhibited by limitations that were originally imposed by hand-calculation methods, and later came to be considered intrinsic to the methods themselves. A typical specific example is the conventional assumption of full rank (nonsingularity) in statistical procedures such as linear regression, and the resulting problems due to near-singularity (multi-collinearity). Once the linear model is described in the natural, general framework of orthogonal bases, the *computational* problems of general rank in linear models are by no means insuperable. To exploit the computational methods fully, however, one must consider the *statistical* implications of such questions as uncertain rank (see Section **5.f**).

The *costs* of computing are more frequently discussed, at least in the narrow sense of computer time or other quantifiable resources. Section **b** gives some general comments on this topic, and the remainder of the book contains discussions of the cost of specific procedures. The extent to which such costs ought to be the *main* concern in data analysis is questionable, however. For very large, routine projects, computing efficiency may be the determining factor. For other situations, particularly in exploratory work, the other criteria discussed in Section **b** are often of greater importance.

b. Program Evaluation

The evaluation of any set of proposed computing tools can be organized around four main headings:

1. usefulness,
2. reliability,
3. cost.
4. convenience.

Each of these provide a framework for planning and assement.

Usefulness. The most important requirement of a program or computing system is that it solve a relevant and important set of problems. If a particular set of problems provided the incentive for the programs, one must first look at this local question of usefulness. The evaluation will require some care. For example, does the program *fully* solve the problem, including provision of data acquisition and display of results where necessary? The new program need not do all this itself, but if it does not, communication between it and other programs for this purpose must be straightforward, and if the other programs do not already exist, the cost of writing them must be included.

If the program was written for a continuing series of problems, it will normally be designed around only the few examples currently available (and frequently only a subset of these). Planners need to consider the scope of the *full* series before restricting the applicability of the program design. If programmer, designer and data analyst are different individuals, it is particularly useful that the first two feed back their understanding of the program's scope to the data analyst *before* writing the program.

The broader aspects of program usefulness are sometimes overlooked. The question here is whether the program will be easily useful with, at most, straightforward modifications for other problems. Is the program designed so that it can be used for related but distinct problems? Is it possible to understand the program and modify it to apply to different problems? Can it be moved reasonably easily to a different environment, possibly on a different computer system?

The reason for emphasizing this consideration is that real gains can be made here. It is frequently possible to solve an immediate problem with little extra effort (sometimes, in fact, with less effort) in such a way that the general, long-term value of the programs is greatly increased. To achieve this the larger context must be kept in mind from an early stage of the planning. The programmer deeply involved in details is less likely to see this context than an overall planner slightly removed from the details. Where planner and programmer are the same person, this suggests that a conscious effort should be made occasionally to step back from the immediate problem and consider the overall strategy. Sections **c** and **d** discuss several aspects of increasing program usefulness.

Reliability. The reliability or accuracy of computations and the cost of alternative methods are more technical questions that can only be answered relative to the current understanding of particular computing problems. Estimation of reliability is a central theme of numerical analysis, which is discussed in Chapter 4 and elsewhere. A closely related concept is *sensitivity,* the change in the result of a computation as a result of a change in the data provided to it. Where data is uncertain, as nearly all observational data must be, such analysis must become particularly important. Except for a few relatively simple techniques, sensitivity analysis is not yet a well-developed tool in scientific data analysis.

Different questions of reliability arise with respect to the integrity of the data and the programs themselves. Statistical techniques for detecting gross errors in data or for reducing their effect on subsequent calculations have been developed for some of the standard analyses (se Section **5.j**). Computational procedures designed to ensure the integrity of data have been included in many data base management systems.

Cost. Cost estimates and cost comparisons have been developed for many algorithms and is discussed frequently in later chapters. There are two basic approaches to cost evaluation: theoretical and empirical. The most desirable measures of cost would seem to be *actual* cost, and for the more complicated procedures no other measures may be available. The relevant cost measures will depend on the machine and the viewpoint of the user. On a small or wholly dedicated machine with a single user the natural measure is the elapsed time required to complete the job. Even in so simple an environment, some qualifications are necessary. The true cost of lengthy computations may be substantially lessened if they do not require constant supervision. On small computers that are normally idle at night, users are often ingenious at inventing computing tasks that can be started and left to run indefinitely. Such programs, it may be argued, run at very little cost, however long they take. (Multiprogram systems, applying similar logic, may provide a low cost, background, or deferred grade of service, which runs essentially by use of otherwise idle resources.)

On modern multiprogrammed systems, cost measures are more indirect. The supplier of the computing facilities wishes to obtain the most computing from the available facilities, by ensuring an efficient allocation of resources among users. In a commercially organized computing facility, the supplier will generally charge users for their consumption of the various resources within the system (central processor time, peripheral storage, input/output, printing, etc.). In a purely commercial environment, this will be done so as to realize a suitable return to the supplier. In a semicommercial situation, as in a university or scientific establishment that runs its own computer center and charges others in the organization for computing, the goals are usually to obtain efficient use of the computer and, perhaps, to distribute the cost of supporting the computer facility among members of the organization in proportion to their real use of it. These latter goals may conflict and the usual marketplace approach to pricing computer facilities does not automatically guarantee efficiency in the use of the facility. Various alternative schemes have been suggested for in-house computing allocation. There is unlikely to be an ideal scheme for all circumstances, but a key ingredient would seem to be responsiveness to actual usage and an awareness of current or potential blockages in the flow of user programs.

The user's strategy in a market or semimarket environment generally is to obtain as much of the computing he wants as possible given his budget for computing resources. As a result, the cost of alternative computational procedures or systems must be evaluated in terms of the totality of resources required. A larger program that executes faster may or may not be preferrable; similarly, the choice between data management that uses more peripheral storage and one that requires more frequent access to that storage will depend on current pricing policies. This is part of the rub from the user's viewpoint. The flexibility in pricing that the supplier sometimes needs to respond to varying demand may make it difficult for an individual user to develop a stable efficient algorithm for some of his major computations.

The topics we have just discussed have become significant technical problems with the development of large, complex, multiuser computing systems. Simply measuring and understanding the performance of such systems and analyzing the results of changes in the systems or their pricing strategy may require sophisticated data analysis. Such techniques have been labeled *compumetrics.* In fact, complex computing systems provide interesting challenges for data analysis: the data is abundant and significant relationships and patterns are likely to exist, but the appropriate models and estimates are frequently not obvious.

In addition to the ambiguities and instabilities mentioned, empirical cost comparisons are obviously defined only for a particular machine and frequently change when the same procedure is used with a different mixture of other programs, particularly in a multiuser environment. Designing or selecting algorithms for multicomputer use requires some measures of cost that are valid generally. For this purpose, theoretical cost calculations may sometimes usefully supplement empirical measurements.

To obtain such results one must be able to estimate some measure of the computation required from a knowledge of the algorithm itself rather than a specific implementation on a specific machine. The method usually employed is to *count* the number of certain special operations required in the execution of the algorithm. In a numerical algorithm, one may be able to count or estimate the number of basic arithmetic operations (add, subtract, multiply, divide), plus possibly the calls to standard functions (square roots, logarithms, exponentials, etc.). In sorting or table look-up one may compute the number of comparisons required. More complicated algorithms may require less direct measures. For example, a model-fitting or integration routine would require a user-written subprogram that the algorithm would call repeatedly. A natural cost estimate is then the number of such calls required to achieve a desired accuracy. For data manipulation on peripheral storage one may be able to estimate the number of accesses to the peripheral device and/or the amount of data transferred.

Given some operation-counting ability, one may compare alternative algorithms for various possible sets of input data. In some situations, (such as for some linear procedures discussed in Chapter 5), the operation count depends only on the *extents* of the input data, such as the number of rows and columns in a matrix. Counts can then be estimated as a function of a few integers and fairly simple formulas derived (see Section **5.g**). More generally, the operation counts depend on the values of the data as well as the extents. Sorting, for example, generally depends on the initial ordering of the input data.

Since quoting costs for all possible inputs will then be impractical, some metric or averaging process must be applied. Suppose, for illustration, we have an algorithm that operates on a single input array, X, of extent N. While the actual cost depends on the value of X, we need a cost estimate only involving N. The two most commonly used measures are the *maximum* cost over all X and the *average* cost on the assumption that X is generated by some random process. Maximum cost is favored by computer scientists in many cases because it provides a guarantee of the algorithm's performance. In some algorithms maximum costs are much larger than average costs and correspond to inputs that are unlikely to occur, in some reasonable sense. Some sorting and table look-up algorithms are examples. Then average costs may be more reasonable. Unfortunately, most computational models of random inputs are rather primitive. The tendency is to assume all the input numbers independent and usually uniformly distributed. More realistic and flexible models for average cost calculations would be useful in many applications. For example, random inputs to a sorting program might be generated over a complete range of internal order dependencies (Chambers, 1971).

The crucial assumption in the use of operation counts as cost estimates is that costs incurred in the algorithms that are not proportional to the operations counted can be ignored. When the operations counted are quite expensive, as often is the case for optimization or integration algorithms, this assumption may be reasonable. Otherwise, one must use experience and occasionally intuition to decide whether the operation counts are a reasonable guide. One may be suspicious of an algorithm promising a significant reduction in operations but appearing much larger or more complicated in logic than its competitor.

A useful extension of ordinary theoretical and empirical cost measurements is the *frequency counting* or *profiling* of the steps in an algorithm. This may be done empirically by invoking a counter routine at important points in an algorithm, with arguments to the counter program that specify the algorithm step from which it was invoked. An empirical profile of the algorithm then emerges as the set of frequencies with which the various steps were executed. if one assigns a time cost to each step and the profile is complete (i.e., all the executable statements fall into one of the counted steps), an overall timing will result. The profiling results are also important in deciding which steps dominate costs and, therefore, deserve of particularly careful coding. Debugging may also be helped if it is noticed that certain steps are executed too often or too seldom.

It is relatively easy to design a profiler that can be invoked on top of a general programming language. For example, a preprocessor could look for comments in a special format. These are translated into calls to the counting program. Then, omitting the preprocessing would delete the profiling (for greater efficiency, once the debugging is complete). An interesting technical problem is to construct an efficient network of counting calls to completely profile a given algorithm. See Knuth and Stevenson (1973) for an analysis. Some available profilers are listed in the Appendix. Clear and useful profiling is much enhanced by good structure in the program design (see Section **c**).

Profiling is a useful tool whose applications seem certain to grow in the near future. A few limitations should be kept in mind. Dollar cost is not always determined uniquely by the profile; for example, the program size may affect its cost. The time taken for complicated programs and, particularly, iterative numerical ones may depend on machine precision or other machine constants. Nevertheless for many programs, profiling provides a very useful and reasonably machine-independent evaluation. For particularly basic algorithms theoretical profiles can be constructed. Knuth's multivolume work, *The Art of Computer Programming,* contains many such analyses, using a fictitious machine of his own devising.

Convenience. The issues of reliability and cost are largely technical. The appropriate measures for specific problems may not be uniquely defined, and more complex problems may make any measure difficult to estimate. Nevertheless, the subjective component is fairly small.

The convenience of use is much less quantifiable. Also, it tends to depend on the overall form of the computing system used, not merely on a few specific operations. Some relevant questions to ask are:

1. What is available? Does the system contain methods to solve the relevant problems?
2. How easy is it to use? Does the system handle unnecessary details, while leaving the user enough flexibility to specify the desired analysis precisely?
3. Can the system grow? Can new methods and new applications be added easily, even if they involve algorithms developed elsewhere?

Answering the first question involves comparing available methods with the current best. This is the purpose of much of the remaining chapters of the book. The second and third questions reflect both the user *language* and the facilities for description and manipulation of *data*. Chapter 3 discusses the latter.

Significant advances have been made in the techniques for describing languages and compiling or interpreting programs written in them. Compiler-generating programs, which construct a compiler or interpreter from a relatively simple and comprehensible description of a language, have become much more powerful and useful. This and other advances make it possible to define or modify languages with considerably less effort and compromise. As such techniques become more widely known, and as some remaining questions on topics such as appropriate data structures and interface to operating systems are solved, the scope and effectiveness of languages for data analysis will increase dramatically. Sections **e** and **f** discuss programming languages in slightly more detail.

c. Program Design and Structure

This section discusses writing good programs. It complements Section **b** which discusses the *evaluation* of programs. We now consider the development process from the inside, and discuss some criteria for *creating* high quality programs. Through most of its history, programming has been essentially a craft, occasionally an art, but rarely a science. As with many crafts, principles have been developed to guide practitioners. The rationale behind the principles tends to be vague at first, but gradually the more important rules develop towards a logically coherent structure. As this process continues, the craft, or at least some part of it, may develop into a scientific discipline. We consider here a few topics in program design. For the many other topics involved in good programming, readers are referred to texts on programming style, such as Kernighan and Plauger (1974). Also recommended is a later book by the same authors (Kernighan and Plauger, 1976) discussing the design and implementation of some of the utility algorithms whose availability can make the writing of software for any application more productive.

An important development of systematic and formal approaches to programming has begun. The central theme of the development is the creation of programs whose execution is understandable and predictable. Various ideas have been developed to further this goal, frequently denoted by the term *structured programming.* An exposition of this important concept should be part of any serious course in computer programming. Among the books discussing the topic, Dahl, Dijkstra and Hoare (1972) and Mills (1973) may be mentioned, since Dijkstra and Mills were instrumental in developing aspects of the topic. We consider here some of the concepts of structured programming that are relevant to programming for data analysis.

An important principle is that programs should have a well-defined hierarchical *structure*. Symbolically, we can express this as follows. Let A be an algorithm. Then at the first level we show explicitly that A consists of N steps, $A_1; A_2; \cdots; A_N$ where A_i is allowed to be a statement in the programming language or a group of statements subject to certain restrictions to be discussed. The effect of each A_i is now defined as precisely as possible, which in turn defines the effect of A. The next stage is to take each of the A_i in turn and apply the same process to it; that is, we describe A_i as a sequence of steps A_{ij}, $j = 1,...,N_i$ and define the effect of the A_{ij}. The process continues until each step is elementary, so that it does not need to be analyzed further. This process of description is called a *top-down* analysis.

Two specific aids to structured program design that are of particular importance are *modularity* and restrictions on *control transfer*.

Modularity derives from an architectural term for the repeated use of standard unit pieces in construction. It is borrowed in computing to refer to the design of programs around procedures or subprograms that perform a well-defined function and are largely independent of the context in which they are used. The term is thus a relative or descriptive one, describing a philosophy and goal rather than a precise technique. Within reason, modularity contributes to the general usefulness of programs, as they can be used in unforeseen applications with little adaptation.

Many aspects of programming style contribute to modularity. Some are quite obvious; others considerably less so. Portions of the program that perform some well-defined calculation, depending on a specific, limited amount of outside information, may be written as separate subprograms. Recognizing the operations that should be separated in this way requires clear understanding of the nature of the computations. The subprograms should be those that may be used, without change, in a number of related calculations; that is, they should be the basic natural units from which algorithms for the subject area can be constructed.

When modules are well chosen, they may produce several additional advantages. Because calls to modules may be used to replace equivalent code in new algorithms, the writing of these algorithms is made easier and less prone to mistakes. The use of a single subprogram instead of repeated equivalent code may reduce the overall size of the program. Particularly important modules may be carefully turned for accuracy, efficiency, elegance of display, or whatever are their important features. The advantages will then automatically extend to all the larger algorithms using the modules. Modularity of the nonportable portions of a programming system (see Section **d**) also assists in transporting the programs to a new environment. Thus portions of code that are dependent on the local machine or operating system should be isolated into as few and as simple a set of modules as possible.

A proposed structuring of algorithms into modules may be evaluated by considering how well they provide the benefits mentioned above. Can the modules in fact be used easily in different computations? It may happen that much auxiliary computation is needed to make either the inputs or the outputs of the module conform to its new context. The modularity of a calculation may be the result of complicating the calling sequence or increasing the number of arguments beyond what a particular application would require. The more trivial the computation, the greater is the need to justify the module by its frequent and important occurrence. A very small module, or one with many or complicated arguments, may lose some of its economic advantage by reason of the cost of linking the calling program and evaluating the arguments to the subprogram for the module.

Modularity has been equated, up to this point, with the use of subprograms for important computational components. Other methods, depending on the programming language, may be used to isolate portions of computation, such as macros and related features (e.g., statement functions in FORTRAN). These provide some of the same convenience in programming, but the resulting code is generated separately for each use of the module, and no linking takes place.

The organization of storage also involves modularity, analogously to the organization of calculations. Values that recur frequently in the computation may be isolated as *data modules.* The precise meaning and the implementation of this idea depend strongly on the programming language used. In FORTRAN, for example, data modules may be implemented as blocks of labeled common. The variables in the block are accessible from any subprogram that contains the common block declaration. Such blocks could be used, for example, to provide machine-dependent constants (relative precision, limits on the size of numbers, relative lengths of different types, etc.).

ALGOL60 does not have a corresponding concept. Data modules would probably be organized as nonlocal arrays. The name of the array is then accessible directly to all procedures declared inside the block containing the declaration of the array. One useful result in both the FORTRAN and ALGOL case is to reduce the number of arguments which must be passed down the sequence of calls. If A calls B and B calls C, then the information required from A by C must all be passed through B if no data modules are used. If B does not need this information, the potential modularity in writing B is seriously reduced. The use of data modularity is thus an important aid to program modularity.

A different approach to data modularity arises when several items or arrays are naturally associated in a data structure. In languages such as PL-1 and ALGOL68 it may be possible to declare a *structure* (see Section **3.c**) that represents all of these items collectively. In communication to subprograms, only the name of the structure is needed, rather than the names of all the component arrays.

Control transfer refers to conditional jumps, looping, and any other mechanism that causes the sequence of computation to depend upon previously computed values. Examples are the DO loops, IF and various GOTO statements in FORTRAN, the **for** and **if** statements in ALGOL60, and similar constructions in other languages. In verifying the correctness of an algorithm, one must consider all possible paths of computation defined by control transfer. Such a complete analysis will be made more difficult if the program contains a complicated, unstructured set of jumps back and forth. (The particular target of criticism is the ordinary GOTO statement found in most programming languages.) The strongest form of analysis is the formal *proof* of correctness (for example, Hoare, 1971). While few data-analytic algorithms are candidates for formal proof, similar arguments may be applied to less formal verification and to ordinary program testing.

The discipline imposed on control transfers might be roughly as follows. We require the executable part of a program unit, A, to consist of *steps*, say

$$A_1;\ A_2;\ \cdots\ A_N;$$

where each A_i is either a simple statement (one which cannot alter the sequence of execution) or one of a set of permissible control statements. Exactly what is permitted remains a matter of debate, but the two essential features are:

1. *iteration*, that is, the repeated execution of a set of statements, with the number of repetitions controlled by some computed quantity;
2. *alternation*, that is, the choice of one of a set of possible computations, on the basis of some computed quantity.

Figure 1 illustrates one way of implementing these concepts.

```
switch: {
  case(expression1): statement1;
  case(expression2); statemtent2;
    ...
  case(expressionn): statementn;
  default: statementdef;
  }
```

Figure 1(a): Alternation

```
repeat {
  statement1; · · · ;statementn;
  }
```

where each statement is either simple or of the form

```
if(expression)break
```

Figure 1(b): Iteration

In 1(a), the expressions are evaluated as logical (boolean)

quantities. The first one that evaluates as *true* causes the corresponding statement to be executed. None of the following expressions is then evaluated. The **default** statement is executed if none of the expressions is true. In 1(b) the loop of statements is evaluated until some statement of the second form causes a **break** to occur. The various statements in Figure 1 may themselves be either simple or control statements of the allowed type.

In Figure 1, we have reduced the possible forms of control statement to two forms, that are quite powerful and therefore not particularly restrictive on the form of control. FORTRAN DO-loops, **for** loops in ALGOL60 and many other constructions are special cases of the form of **repeat** loop given. The FORTRAN and ALGOL **if** statements are special cases of the **switch** statement. Some analyses of program flow would be easier if the form of statement were further simplified. Even so, programing entirely within the structures given, and their special cases, would give programs greater structure than unrestricted programming in, for example, FORTRAN.

The goal of structured programming is to produce programs that are understandable and whose correctness can be verified, as far as possible. The specific forms of control structure permitted, or even whether there should be absolute restrictions of this form (Knuth, 1974), are discussions of the *means* to attain these goals.

d. Portability of Programs

The *portability* of computing software (i.e., the ability to use the same or very similar software on more than one computer system) contributes to long-range usefulness for many applications. The achievement of reasonably portable code depends on a number of factors, some more easily controlled by the programmer than others:

1. writing programs in a portable language;
2. structuring programs so that unavoidable machine dependencies are isolated in a few program or data modules;
3. structuring numerical calculations to keep dependencies on precision and range (cf. Section **4.b**) modular in the same sense;

4. organizing data storage and management to minimize dependencies on the physical storage media, character code, or other local features;
5. writing programs for graphical and other output so that most of the code is independent of the particular device (see Chapter 8).

The most stringent approach to portability requires the actual programming to be directly usable on all machines of interest for the project. Two large efforts in this spirit were the ALTRAN system for symbolic algebra (Brown, 1976; Hall, 1972) and the PORT library for numerical analysis (Fox et al., 1976). See the Appendix and Ryder (1974). This approach becomes increasingly difficult, however, when such nonnumerical areas as data management and graphics are involved. (Some relevant points in these areas are discussed in Chapters 3 and 8.) For data analysis applications, therefore, some more flexible approach is needed. One such is to assume that algorithms are written in a language and format that are in principle machine independent, but that may not be directly usable on a given machine: for example, in an extended version of FORTRAN. The algorithms are then preprocessed as necessary, for a *given* machine by a "precompiler", whose output is, in this case, in a dialect of FORTRAN acceptable to the given machine, but possibly containing machine dependencies. Much work on this approach has been done; the graphics system described in Section **8.a** uses this approach. Cook (1976) describes a representative example of the approach. Current procedures tend to attack the various portability problems piecemeal, and either leave a great deal of work to the local implementer or else require the program writer to do much more work than in writing non-portable code. The long-term solution seems likely to be in the context of designing new, FORTRAN-based programming languages.

e. Programming Languages: General Discussion

As computing technology developed in the 1930s and 1940s, the techniques for causing a sequence of calculations to occur increased in flexibility and complexity. Mechanical and electrical calculators had cranks, buttons or levers to initiate single numerical operations. The sequence controlled calculators of the early 1940s, borrowing the principle developed over a century earlier by Jacquard for mechanical weaving and proposed by Babbage for computing, specified a sequence of such operations by codes on paper tape or cards. The calculator could then carry out a number of operations, repeatedly using and changing values held in internal registers, without intermediate interference by an operator.

The great innovation, and perhaps the best boundary between calculator and modern computer, was the concept that the operation codes could also be stored in internal registers and modified as necessary during the calculations, registers holding operation codes being essentially indistinguishable from those holding data. This was an intellectual breakthrough of enormous significance, opening up such basic computing techniques as indexing and indirect addressing, iterative computations, subprograms and data structures. The first stored program calculators (e.g., EDSAC) were primitive and gargantuan in hardware, but are nevertheless recognizable computers in the modern sense.

These brief historical comments may appear rather remote from our topic, but they are essential to give a proper perspective to discussions of modern computing. It is frequently necessary to emphasize the revolution in concepts about calculation, deriving to a considerable extent from the stored program concept. Many presentations of scientific techniques were devised for the precomputer world, and a number of these have been made entirely or partially obsolete. As is usual, what is traditional becomes thought of as natural, and the generation or so since the development of computers has been barely enough to begin the revision of outmoded concepts.

In the narrower context of user communication, a historical perspective is essential to understanding why programming languages developed and what they do. The stored program soon led to the desire for much larger and more complicated sequences of calculations. A justifiably famous early example was the von Neumann and Goldstine (1947) procedure for matrix inversion. In order to simplify the definition of the computations, the idea of coding the operations indirectly, rather than in the numerical codes to be stored in the machine, developed around 1950. The codes used have developed into what are now known as programming languages. For an extensive review of the earlier history, see Knuth and Pardo (1976).

Initial work on languages was largely on a case-by-case basis. Experience in designing languages and work on related theoretical ideas has gradually produced general methods for defining and implementing new languages. For those involved in major computing efforts for data analysis, the important implication is that communication between the user and the underlying algorithms should not be regarded as imposed from without by existing languages or systems.

A programming language may be considered as an algorithm that processes strings of characters (the program, or pieces of it), attempts to match patterns in the strings (according to the *syntax* of the language), and takes a sequence of actions (according to the *semantics* of the language) accordingly. A key advance in language design has come from being able to define the language in a formal, symbolic way. The formal definition describes the patterns recognized in the language and, optionally, the corresponding actions. This description is given to a *compiler generator* that produces as output the algorithm for processing the given language. Because the compiler generator need only be written once for the class of languages it recognizes, much of the important, difficult, and delicate work in producing the language is taken over from the designer of a specific language. In using such a compiler generator, new languages usually can and should be defined by extending corresponding existing language definitions.

Readable discussions of some principles of language design and translation are given by Johnson (1976) and Aho and Johnson (1974). For readers wanting a fairly deep understanding of modern techniques in this area, Marcotty et al. (1976) is worth careful reading. The most influential example of language design in recent years is probably ALGOL68. An introduction to the language and its principles is given by Tannenbaum (1976); it is recommended reading for anyone planning a new language.

f. Programming Languages: Comparisons

We turn next to a discussion of some programming languages currently in use that may be candidates for applications in data analysis. Programming languages facilitate three aspects of computing:

1. The organization of data within a program and its transfer to and from external storage media;
2. The control and sequencing of calculations on numerical and other data;
3. The presentation of computational results and other information.

Programming languages differ greatly in the approach taken to each of these areas, in terms of their scope, complexity and logical organization. Their history is complicated and fascinating. A thorough understanding would involve many viewpoints: the informal approach to planning most early and some later languages; economic considerations for sellers and user; the sociology and psychology of language use; and the attempt to develop a precise, logical basis for language development. No such treatment of the subject is attempted here. Many of the aspects noted have not been fully explored as yet. For an extensive historical summary of over 100 languages, see Sammet (1969).

	FORTRAN	BASIC	PL-1	APL	ALGOL60	ALGOL68
Availability	Good	Good	Good	Fair	Fair	Poor
Machine Efficiency	Good	Fair	Fair	Poor	Poor	Good
Structured Programming	Poor	Poor	Fair	Poor	Fair	Good
Data Structures	Fair	Fair	Good	Good	Fair	Good
Existing Algorithms	Good	Fair	Fair	Fair	Good	Poor
Interactive Use	Fair	Good	Fair	Good	Poor	Poor
Ease of Initial Use	Poor	Good	Poor	Good	Poor	Poor
Use on Small Computers	Fair	Good	Poor	Good	Poor	Poor

Figure 2: Comparison of Programming Languages

Figure 2 is a subjective scoring by the author of six general-purpose programming languages on a number of components that may be of importance to various projects. The scoring is *strictly* from the viewpoint of data analysis; i.e., no implications should be drawn as to the suitability for very different areas of computing. The remainder of this section discusses the figure in a little more detail. First, however, we should consider the alternatives not included in the evaluation, specifically, languages not suitable for data analysis, special-purpose programming systems, and dialect variations. COBOL and various assembly languages make up the primary members of the first class. COBOL is still widely used for various business applications, and has some desirable features in the areas of data structure and data base management (see Chapter 3). However, its computational and logical facilities are extremely cumbersome, and it is not a natural medium for most data analysis. The benefits of efficiency in assembly language coding are not sufficient to offset the clumsiness of program writing, lack of portability, and proneness to error. Some small amount of assembly code is often required to generate a local version of a large system written in a higher-level language (for example, pointer manipulation or character operations in FORTRAN).

The special-purpose systems and languages are excluded mostly because they are more of an alternative to the use of general languages rather than another choice among them. Some of the systems are quite effective aids to data analysis, particularly for the casual user. In the more sophisticated, the user may indeed consider writing programs or even algorithms. With respect to the algorithms discussed in this book, however, one will likely either find some version of the algorithm in the system, or else will consider adding the algorithm to the system, in which case the choice of language is already made. Some considerations with respect to statistical packages are presented in Section **1.b**. Two of the languages described in Figure 2 (BASIC and APL) are more directly comparable to packages, in that they are intended for ease of initial use, particularly on simple problems.

The languages mentioned in Figure 2 are intended to represent "common" versions. For most of them, a number of dialects, extensions or subsets also exist. A prospective user should consider whether the local version of a language differs sufficiently either to overcome some of the disadvantages or to compromise the general availability and portability of programs.

Let us now consider the headings in Figure 2 in a little more detail.

Availability. This is taken to mean, roughly, the chance that a typical data analyst will have convenient access to the language. FORTRAN is clearly dominant, with the exception of use on smaller machines. BASIC and PL-I are about equally available, but not together: the former appears on interactive systems and small machines, the latter mostly on medium-to-large systems. At this time, ALGOL68 has very limited availability, but ease of implementation was a design goal for the language, and increased demand may bring greater availability.

Machine efficiency. This is not to be confused with overall efficiency, which includes various human costs (see Section **b**). The efficiencies of PL-I and ALGOL68 are potential, but depend on knowing the languages and their compiled code in some detail. APL is reasonably efficient for simple operations, but sometimes spectacularly costly for complex programs.

Structured programming. This is the degree to which the language facilitates the kind of program design discussed previously. It should be noted that various combinations of programmer discipline and/or preprocessing may simulate at least the control transfer aspects in the poorer languages (see Section **d**).

Data Structures. ALGOL68 and PL-I are rated high for the ability to declare structures of a general kind (see Section **3.c**). APL is rated high for an entirely different reason: it is the only language with a completely general, natural approach to arrays of an arbitrary number of subscripts. It has *no* facility for structure declaration.

Existing algorithms. This is the combined question of what has been written in the language and how easy it is to find. Notice that some systems have interlanguage interface capabilities (for example, from PL-I or ALGOL60 to FORTRAN). If these are really adequate, the availability of algorithms is of course much improved.

Interactive use. This is interpreted in a narrow sense of more or less conversational interaction, as opposed to generating jobs to run in batch from a terminal.

Ease of initial use. Here it is well to consider the special-purpose packages and languages as competitors. The best of them are probably superior to any of the general languages on this component, for data analysis use.

Use on small machines. The term "small" is most relevantly measured by financial investment. The same languages are rated high here as in the previous component, reflecting perhaps their simplicity or their stress on the casual user.

Some tentative conclusions emerge. For the serious writer of algorithms, who wants widespread availability of the results,

FORTRAN remains the best choice, particularly if reinforced by extensions or pre-processors as discussed in Sections **d** and **e**. At the time of writing, the Revised Standard FORTRAN has been presented but not adopted. It moves FORTRAN ahead somewhat on program structure and data structure, but otherwise it does not imply a major change in our analysis. For casual users, BASIC and APL have strong appeal, subject to some extent to the competition of special-purpose packages and languages.

Looking to the more distant future, the many attractive features of ALGOL68 may or may not be sufficient to ensure its widespread availability. A more relevant hope, however, is that continued advances in the art of language design and implementation will allow us to develop and extend languages to suit our own needs, obtaining a good approximation to the best of all possible worlds with respect to language design.

Problems.

1. Design a program which copies the text of a given program and inserts statements to count each occurence of specified function/subroutine calls or arithmetic operations. Try it first ignoring, then including conditional expressions. Hints: (i) A completely general version would require the ability to *parse* FORTRAN statements, that is, to determine the type of statement and, at least for conditional statements, extract the pieces of the statement. (ii) It will help to build the algorithm on primitive routines to read a statement and return its type, to find and delete an occurrence of a string of characters (Kernighan and Plauger, 1976, Chapters 5 and 6), and to write out a statement.

2. One problem in making FORTRAN programs portable is that the language has no concept of character-string variables.

(a) Design a preprocessor that converts declarations of character-string variables, say

CHARACTER(name,length,initial.text)

into a declaration of the named variable as an integer array that can hold *length* characters and has the given initial text. How could this be extended to allow the declaration of multi-way arrays of character strings?

(b) Design operations to assign values to such variables and to extract substrings. First assume characters are stored one per word; then assume the text is packed and elementary routines

```
jget(string,i)
jput(string,i,character)
```

exist to retrieve and store the ith character.

The facilities described could be implemented either as functions or macros in a suitable string or macro language, or by defining the syntax of an extended language with FORTRAN as output. The latter is more ambitious, but gives a more useful result.

3. Design a language that acts as a pre-processor to FORTRAN and implements the control structures of Figure 1. First, specify what FORTRAN should correspond to the specified control structures. Next, consider how this would be implemented. See Kernighan and Plauger (1976, Chapter 9) for a detailed discussion. How would you include the CHARACTER facilities of the previous problem as an addition?

CHAPTER THREE

Data Management and Manipulation

This chapter describes some of the techniques of data management and associated topics, and relates them to data analysis. Techniques for data base management are discussed extensively in the general computing literature, but the terminology and typical examples are often unfamiliar. The objectives and constraints for data management in data analysis are also significantly different from commercial applications. An approach to data management is important, however, both because statistical data must itself be managed and because analysis tends increasingly to operate on data bases already set up for other purposes.

Sections **a** to **c** discuss techniques for managing storage space and organizing it into vectors, arrays or general structures. Section **c** includes a brief introduction to the topic of data base management systems. Sections **d** and **e** present methods for sorting data and searching in a table. Section **f** gives some general comments on statistical use of data base software.

a. The Management of Storage Space

A program running in a computer system has available to it various media for the storage and retrieval of data. Available space may be *internal* in that it is directly referenced in the programming language through arrays or other data structures. Other media may be *external,* in that reference to them is by input, output or control operations executed by the program. In simple systems, the user's program resides entirely in the main (core) memory, and all internal references to data are equivalent. Operating systems with *virtual memory,* however, may partition a user's program into segments or *pages,* and allow the system to swap pages between main memory and an external medium. While the user is oblivious to this distinction, in principle, frequent page swapping carries considerable overhead. For example, some array operations may become markedly less efficient if paging is ignored (see Section **b**).

External media, in turn, will be treated differently, depending upon their physical characteristics. Most external storage management takes place on a *direct access* medium (usually magnetic disk) for which the speed of data transmission to or from any part of the device is approximately constant. A computer operating system may provide relatively high-level storage management (including some or all of the facilities described in this section). If facilities are lacking, or if the user wishes to take greater control over the methods used, algorithms may be written to manage storage directly.

Problems of storage management occur in different contexts and with different objectives. Operating systems may allocate both internal and external storage media among multiple users, either automatically in a multiprogram system, on demand by user programs or for the system's own changing storage requirements. Some programming languages provide for changing storage requirements within a single program.

The important point in intelligent design or use of data management facilities is to understand what basic techniques are available, how they differ, and what considerations help to determine the appropriate technique for a given application.

We first consider the standard forms of storage allocation. A subset of these are characterized by allocating and releasing storage from a conceptually contiguous set of items. Some such storage procedures are shown schematically in Figure 1. The contiguous set of items grows and shrinks as items are added and removed, the distinctions among the forms being whether additions and deletions come at the same end. The most common dynamic allocation algorithm is probably the stack, or the *last-in-first-out* procedure. This is the form of storage provided by ALGOL60, for example, and it is usually the appropriate form for "scratch" storage in numerical algorithms.

In a language, such as FORTRAN, that does not provide stack storage, an explicit equivalent can be provided. An array, WK, of working space is provided in a COMMON block. Programs wishing to obtain space call an allocating routine, requesting a block of N locations in WK. The actual allocation is generally N+1 locations, one of which gives the length of the block. The allocating routine returns an index, say I, for the block. Then WK(I) through WK(I+N−1) are available as the block. A second routine is called to release storage explicitly, using the stored length to remove the block. The allocating and releasing routines must know the index of

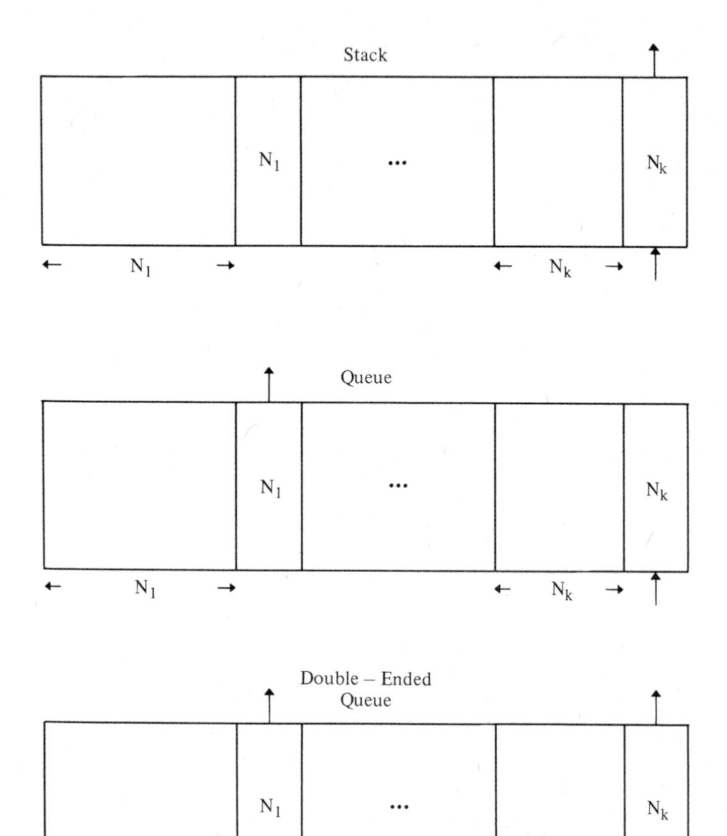

Figure 1: Storage Structures

the end of the last block allocated and the size of WK. Fox, Hall and Schryer (1975) describe a FORTRAN system on these lines.

A disadvantage of this approach is that the total amount of working space must be specified when the programs are compiled. However, by sacrificing some degree of portability, a scheme may be developed in which the actual working space limits can be changed from run to run. It may be possible to find the location of unused memory, in a single program system, and allocate space there. In multiprogram systems, larger blocks of space may be obtained from the operating system and subdivided for working space as described.

Queues and double-ended queues arise in operating system applications, commercial programming, list processing and many other areas. They are perhaps less frequently relevant for data analysis than stack storage, but are similar in their programming requirements. The essential change is to regard our available space as circular. Now both top and bottom of the occupied space may move, but adding or deleting from either end is still very simple. To implement circular addressing, we compute all our indices modulo the length, L, of our working area.

The next level of complication arises when the blocks to be allocated are not necessarily contiguous. Notice that the top item in each block previously served two purposes. It gave the length of the current block and told how to find the one beneath. These two functions must now be separated. It is usual to say that the blocks are *chained* together into a *list.* Such a form would be required, for example, if several queues or stacks were to share the same working area. An attempt to implement allocation and release in this form will soon expose one further complication: where does the space for new blocks come from? The usual technique is to define one more list in the working area: the list of *free space.* These are blocks not currently used in any list or queue. When a new block is to be put into a list, the free space list is searched for a large enough block, from which the necessary space is taken.

For most data-analytic applications, the critical distinction between stack storage and list storage is that the latter imposes no constraints on the order in which allocated blocks are released. With stack storage, the rule that the last block allocated is the first released must be adhered to absolutely, or the entire storage scheme will collapse. The greater flexibility of the list storage scheme is paid for by extra overhead. In addition to the extra space required in the chain, allocation and deallocation require more operations. If the available free space is not to become so fragmented as to be useless, it may be necessary to reorganize the entire work space occasionally.

This operation, usually called *garbage collection,* consists of moving allocated blocks around so that the free storage area ends up as one contiguous block. Notice that the user of a storage system with garbage collection cannot safely refer to the actual address of a block. Rather, he operates on a pointer, or **ref** in ALGOL68 terminology, that directs him to the block. For example, the pointer might be a variable in a fixed location whose value is the current starting point of the block relative to a chosen base location. Since a block may have been moved since the last access, each new access is a two-step process at least. When the look-up requires, say, access to back-up storage, the extra time required may be relevant. Section **e** discusses techniques for look-up.

We have noted that free-list technique of storage allocation is used when more than one list is to be maintained. In fact, the most extreme example is also a common and useful one, namely, when each "list" is just one block long. The storage system is then called CONTROLLED in PL-1 or **heap** in ALGOL68. In effect, it gives out and takes back blocks of storage on demand, with no constraints on the ordering of allocation and release. If the allocation is done on demand, the procedure requesting the storage implicitly assumes responsibility for releasing it when it is no longer required. If the allocation takes place in the natural execution of a programming language, some effort may be required to determine whether there are any existing **refs** to a given block.

Routines to carry out the requirements of list-type storage allocation may be written in FORTRAN using an array of working space. There have been a number of such schemes: Bray (1974) describes and lists some FORTRAN routines for a general list structure. For general discussions of storage allocation on demand see Section 2.5 of Knuth (1968) or the paper by Baecker (1970) for greater detail.

b. Arrays and their indexing

A common process in any computing is to apply an operation in a loop, indexing over one or more sets of numbers involved in the operation. It is natural to represent the sets of numbers in programming languages as *arrays*, with a symbol standing for the entire set and a means of denoting a particular member of the set. The basic implementation consists of a base location for the array to which an integer is added to define a particular element. Modern computers have hardware and instructions designed to facilitate this form of addressing.

Scientific computing soon leads us to want arrays indexed by other methods. The most common such extension is to *matrices* or two-way arrays, where two integers are used to specify, conceptually, the row and column of the matrix. The natural generalization of this is to k-way arrays, for arbitrary k, with the element indexed by k integers.

The notation used to refer to arrays varies somewhat from one programming language to another. For our purposes, let individual array elements be denoted x[i], Y[i,j] and so on.

Virtually all programming languages index an array by the same basic scheme, with two minor variations. A one-way array is a contiguous set of elements indexed by one integer. A two-way array is a contiguous set of one-way arrays of equal size, with one integer to index the subarrays and the other to index within the subarrays. By extension, a k-way array is a contiguous set of (k-1)-way arrays.

One point of difference is which index applies to what. One choice (adopted by FORTRAN) uses the last integer to index the (k - 1)-way arrays; the other (ALGOL,PL-1) uses the first integer. The conceptual difference is minor, altering the determination of locations in an obvious way. The ALGOL order is essentially *lexical,* that of FORTRAN *inverse lexical.* Languages also differ in whether they assume indexing from a fixed origin (e.g., 1 in FORTRAN) or allow programs to declare the origin. The difference in implementation is minor; to simplify our discussion, we assume indexing from 1 here.

Suppose A is a k-way array. If the indices i_j vary from, say, 1 to m_j for $j = 1, \cdots ,k$, then the element $A[i_1, i_2, \cdots , i_k]$ has the same location relative to the first element of the array as the element a[i] has to the first element of a *one-way* array indexed from 1 to $N = \prod m_j$, with the correspondence,

$$i = i_1 + \sum_{j=2}^{k} (i_j - 1) \prod_{r<j} m_r \tag{1}$$

for inverse lexical and

$$i = \sum_{j=1}^{k-1} (i_j - 1) \prod_{r>j} m_r + i_k \tag{2}$$

for lexical.

The simplicity of array storage is in part due to its *linearity.* If the entire array is contiguously stored then many useful subsets of elements are linear sequences; that is, the sequences of subscripts in the corresponding one-way array are of the form

$$i_\omega, \quad i_\omega + \delta, \ldots, i_\omega + (r-1)\delta. \tag{3}$$

We call any such sequence of elements a *section* of the array A. One particular form of section is obtained by fixing all but one subscript and allowing the remaining subscript, say the jth, to vary over all or part of its range. This produces a sequence of the form (3) with $\delta = \delta_j$ where

$$\delta_j = \prod_{r<j} m_r$$

for inverse lexical order and

$$\delta_j = \prod_{r>j} m_r$$

for lexical. A generalization of this (in fact, the most general section) is defined by setting an increment α_j (positive, negative, or zero) for *each* subscript. Then a sequence as in (3) is obtained with

$$\delta = \sum \alpha_j \delta_j.$$

As an example, the diagonal terms of a matrix, A[1,1], A[2,2], · · · , A[k,k], form a section with $\delta = m_2 + 1$ for lexical or $\delta = 1 + m_1$ for inverse lexical.

The access of array sections and their use in computation can therefore be reduced to a simple incrementing of addresses. In simpler languages, such as standard FORTRAN, the user may need to provide the equivalence. This may be done by writing basic routines to apply explicitly to arbitrary array sections. For example, an inner-product function could be written in the form

$$\text{dot}(a, i_a, \delta_a, b, i_b, \delta_b, n)$$

to compute

$$\sum_{k=0}^{n-1} a[i_a + k\,\delta_a]\, b[i_b + k\,\delta_b].$$

In other contexts, the specification of array sections may be contained in a language definition, for example, the IVTRAN specification in Millstein (1973). Between the two extremes, there are languages for which array sections are not a built-in concept but that do not have the equivalencing potential implied by FORTRAN-style array reference. If efficient processing of array sections is important in such cases, the user may need to take over some of the indexing operations from the system.

For the processing of large amounts of data, the economies of linear indexing are important. At the same time, large data structures frequently must be considered as several separate blocks rather than as a single contiguous array. Linear indexing can be applied in this case by computing the block index and the offset in the block. For example, with blocks of fixed length, B, the increment δ changes an offset of ω in block b to an offset of $\mathrm{mod}(\omega + \delta, B)$ in block $b + (\omega + \delta)/B$, so that linear incrementing goes on as before until the offset is no longer within the current block.

Efficient linear indexing will be helped if δ is small compared to B, since the accessing of new blocks will be relatively infrequent.

If $|\delta| > B$, every data item will force a block access and, on some computers, the resulting processing may be very slow. This problem may become serious enough to require copying of the data or choosing an algorithm which accesses the data in a different order (compare the different decompositions in Section **5.c**; some access a matrix by columns and some by rows).

c. Data Structures; Data Base Management.

Any discussion of data structures is influenced by the intended applications. Arrays are introduced first in our discussion since they are natural for much standard data analysis, have significant economic advantages in handling large bodies of data, and are basic to most numerically-oriented programming languages. In some languages, such as FORTRAN and APL, they are essentially the only built-in data structures. From other viewpoints, arrays are perhaps less basic than structures such as lists, trees, and graphs. Knuth (1968, Chapter 2), in an elementary but extensive treatment oriented towards general information processing, actually treats arrays as a special form of list, called an *orthogonal list*.

The use of some data structures that are not linear, in the sense of the previous section, is supported by many contemporary programming languages and is often easy to simulate in others. Where these structures are a natural analog to the data involved, they can be of real assistance to the data analyst. (As with other computing techniques, there is a tendency to apply sophisticated data structures where they give only an appearance of elegance and generality, while actually complicating the analysis.)

A complete discussion of data structures would be too lengthy for inclusion here, and many of the structures are still of marginal relevance to data analysis. The discussion by Knuth gives a readable account of many basic features. Our discussion centers on the structures most appropriate for data analysis and on the special considerations imposed by that application.

The archetype of the data structures we wish to consider is the *list,* defined recursively as follows.

> A *list* L is an ordered sequence of items $(l_1, \cdots, l_k)$ for $k \geqslant 0$ such that l_j is either an atom or a list.

The concept of atom is used to mean anything that does not itself have any internal structure. The programming language LISP (M.I.T., 1962) gave a rather pure implementation of this concept with atoms defined as strings of characters (or in later versions as integers or real numbers). Such an implementation usually delights the logician more than the data analyst, for whom the list structure tends to be overly cumbersome and indirect. An essential feature of lists, however, is common to most data structures and allows us to treat these in a general way, namely, by *grouping* data structures together into a higher-level structure.

First, assume we can create basic data structures such as single numbers of type real, integer, etc.; vectors; arrays; character strings; etc. Then we define, recursively, a *multilevel* data structure as follows:

> A *multi-level data structure* D is an ordered sequence of items $(d_1, \cdots, d_k)$ for $k \geqslant 0$ such that d_j is either a basic data structure or a multilevel data structure.

Programming languages may provide facilities for declaring structures similar to our definition. The PL-1 structure declaration and the ALGOL68 **struct** declaration are examples of such facilities. In such declarations the structure is defined as a list of basic data types and/or substructures. The substructures may be defined explicitly by listing all *their* elements, down to the levels at which all elements are basic data types. Alternatively, mechanisms exist for defining the substructure by relating it to another declaration. The following examples illustrate the process of definition. First, a structure *series* consisting of a character-string, called a *name,* and a one-way real array, called *values,* is defined in PL-1.

```
DECLARE  1 SERIES                                    (4)
                     2 NAME CHARACTER (8)
                     2 VALUES REAL (50)
```

and in ALGOL68

```
        mode series =  struct (                              (5)
                           [1:8] char name,
                           [1:50] real values)
```

(In this and other cases, we give parallel but not necessarily equivalent examples from different languages. The PL-1 version defines and allocates space for an actual data structure. The ALGOL68 is merely a **mode**, that is, a scheme to define *how* to execute an actual declaration.)

Now suppose a number of the individual structures in (5) are to be associated together to form a *data matrix*. There are many ways of doing this, suitable for various applications. Consider the following:

```
      mode dmbyvar =  struct (                               (6)
                          [1:8] char name,
                          [1:50,1:8] char unitname,
                          [1:10] struct (
                            [1:8] char name,
                            [1:50] real values)  variates)
```

This moderately large mouthful says that *dmbyvar* (short for "data matrix stored by variables") consists of three pieces: an 8-character *name,* a 50-by-8 character *unitname* and an array of 10 structures called *variates* defined as shown. Since the lower-level structure was previously defined as *series,* (6) can be reduced to

```
      mode dmbyvar =  struct (                               (7)
                       [1:8] char name,
                       [1:50,1:8] char unitname,
                       [1:10]series variates)
```

In (7), the structure contains all its variates internally. It is usually more flexible to imagine the individual variates as existing independently. In this case, the data matrix contains a set of *pointers* to the variates. In (7), one replaces [1:10]series by [1:10]**ref** series. A third and more familiar alternative is to store an actual matrix of data values in the structure, for example, by using

[1:10,1:50]**real** data in place of the last line in (7). The last two alternatives offer a trade-off between flexibility and ease of interface to conventional statistical algorithms, typical of questions arising in the design of statistical systems. Statistical systems concerned more with interface to existing algorithms will be more likely to store data, at least optionally, in "standard" forms such as the third alternative. Some further comments on these question are made in Section **f**, but a full treatment is far beyond our scope, and many of the decisions depend strongly on the particular application.

Another limitation in (7) is that the array bounds are fixed at compilation. Depending on the programming language used and the implementation of the data structure, various mechanisms are available to make array bounds variable. Although the structure declarations were presented here as simple declarations, it is currently more likely that some special-purpose software will be written to emulate such structures in a language, such as FORTRAN, which does not sufficiently support them.

Data Base Management. A related class of data structure systems are widely used in accounting, inventory management and related computing applications. These are generally called *data-base management systems* (DBMS) and are of many forms. The CODASYL (1971) report is an important attempt at standardization; also, most computer manufacturers and many other groups have more or less sophisticated versions of the basic ideas. Date (1975) gives a general introduction. What we have called basic data structures correspond to *records* in CODASYL, that is, some basic pieces of data whose internal structure is given. Many of the DBMS are rather restrictive in defining record formats; for example, records that are vectors often must have a fixed length. This raises certain problems in data-analytic applications. Depending on the DBMS in question, a record may be either a structure, somewhat of the PL-1 type in (4), or some more restricted data type.

The DBMS then defines a hierarchical or directed structure among record types. The user declares to the DBMS that there exists a directed relationship *from* one record type *to* another: in the usual anthropomorphic style this is sometimes expressed by saying that the first is a *parent, owner,* or some such, of the second. The language developed in biological classification is also sometimes used, calling the first record type the *root* of the second, or the second a *branch* of the first. The relationships are often conveniently expressed graphically as, for example, in Figure 2.

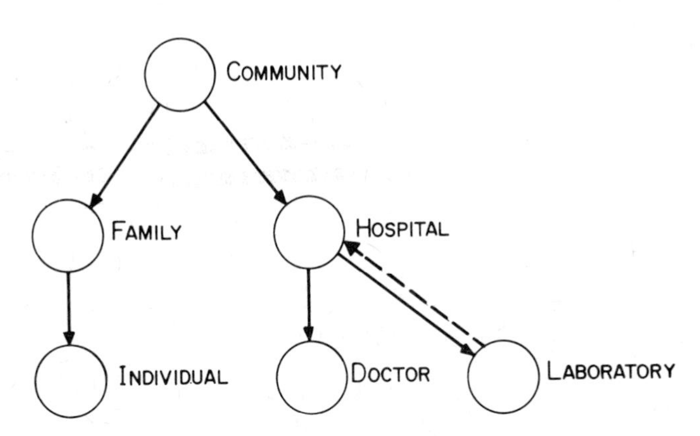

Figure 2: Hierarchical Structure

In Figure 2 we suppose that medical statistics are being collected, say, concerning hospital treatment and its financing. Consider first the solid lines only. This illustrates one way in which a tree structure can be imposed to facilitate analysis of such data. On a high level the data is organized by community and certain summary information (income level, population, etc.) is stored. Data at the next level may be associated with one of several subcategories. The data describing, for example, a particular hospital "belongs to" one specific member of the community category and similarly at lower levels in the structure.

Tree structures are constrained to be strictly hierarchical, in that a category at one level may belong to only one category at the next higher level (and must belong to one, unless it is the single category at the highest level). It may be more convenient to regard a category as belonging to several higher-level categories. In the hypothetical example of Figure 2, including the dotted-line relationship shown produces a *network* or *directed-graph* structure, in which one laboratory may "belong" to several hospitals.

Various DBMS differ in the generality of structural relations allowed and in the terminology used. Usually hierarchical relationships (the arrows in Figure 2) represent inferred properties *among* data sets. Thus the data for a particular community does not physically contain all the data for families, hospitals, etc., which belong to it. Rather there will exist some system of explicit or inferred pointers which are contained in a particular instance of the "community" data structure and which direct the DBMS to, for example, all the "family" data structures which belong to the given "community" structure.

In many DBMS, a major goal is to keep the user unconcerned with the actual representation of the data (usually called *data independence*). In the class of *relational* systems, this includes submerging or eliminating hierarchical or network structure. The user specifies only sets of associated variables, for example, {COMMUNITY, HOSPITAL, DOCTOR}. Such systems are appealing in the abstract, particularly for informal, interactive query programs. So far, implementation difficulties, particularly efficiency, have prevented widespread use of relational systems. Date (1975, Part 2) gives a discussion of the relevant concepts.

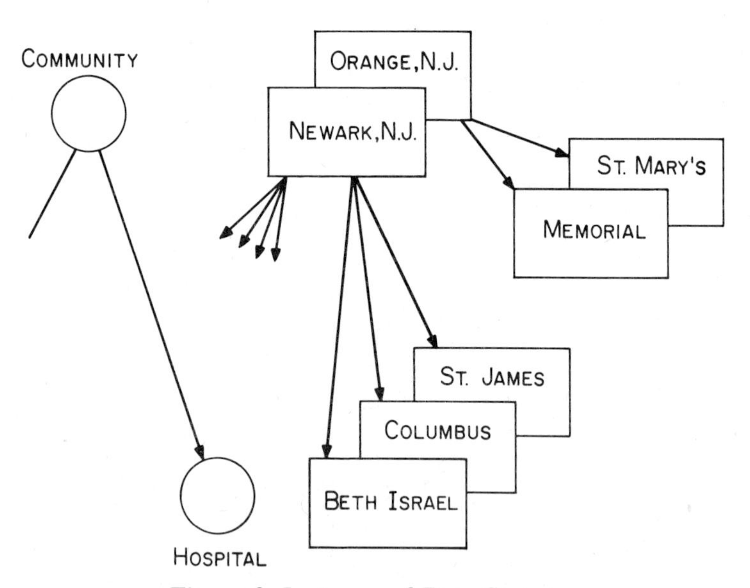

Figure 3: Instance of Data Structure

Terminology in describing relationships like that in Figure 2 can be misleading. If we take *structure* to be defined as above, then relationships in Figure 2 are between types of structures. That is, any single arrow between nodes in Figure 2 implies that a structure of the upper type may point to any number (0 or more) of instances of the lower structure type. Thus Figure 2 does *not* describe hierarchies among actual data. These involve an added dimension: Figure 3 illustrates how a small part of Figure 2 is mapped onto actual data structures.

While many extensive data bases are naturally kept in the systems described in this section, the programming languages and packages most often used in data analysis do not provide for such structures. In addition, statistical algorithms usually assume regular data in the sense of vectors or arrays. When one must reconcile data suitable to multilevel structures with such algorithms, the choice is either to rewrite the algorithm for the data or else to write procedures to copy the data into the required regular form. The latter usually is more practical, particularly when access to structured data is only possible via a special DBMS.

d. Order Statistics: Sorting and Partial Sorting

The general ordering or *sorting* problem is: given a set, say X, of N elements, find a permutation, X_*, of the elements of X such that $X_*[i] \leqslant X_*[j]$ whenever $i \leqslant j$. The relation '$\leqslant$' may be any ordering relationship (e.g., numerical ordering of integers or reals, or lexical ordering of character strings). Using statistical terminology, we call X_* the *order statistics* of X:

$$X_*[1] \leqslant X_*[2] \cdots \leqslant X_*[N] \tag{8}$$

Our discussion focuses on the case of *internal* sorting, meaning that we regard the set, X, as small enough that all of its elements are equally accessible. In other words, X is assumed to be in directly addressable main storage. The contrasting case of *external* sorting assumes that the data must be held on one or more backup storage media (disk or tape). External sorting is inherently a more complicated problem, as one must consider the relative costs of access to different storage media as well as specific limitations on how various devices can be manipulated. The case that is most frequently considered is that of input data on one or more magnetic tapes with some additional tapes available for intermediate storage. This problem is probably farthest from internal sorting in philosophy, since the tapes are generally read straight in with no random addressing capability. Intermediate cases, in which the data was stored on a relatively slow, but random access device, have received less attention.

Further, external sorting is naturally a more time-consuming, expensive operation, in which the details of algorithms are important and frequently not easily changed by a user. For this reason, really large-scale sorting should generally be done by a specialized program, preferably worked out by experts in the topic. A number of such programs have been developed, largely for commercial applications. On computing systems used fairly extensively for commercial data processing, or on any reasonably large or sophisticated computing system, one should look hard for a good existing external sorting program before trying to write one.

A third excuse for concentrating on internal sorting is that most exploratory data analysis tends to be done, almost by definition, on relatively manageable bodies of data. One may argue that an attempt to do extensive analysis of very large data sets is unwise, since the iterative, heuristic manipulation inherent in exploratory work will be restricted to some extent by the size of the data.

We turn, therefore, to internal sorting for most of the remaining discussion. The number of methods that are possibly competitive for this problem is still quite large. We concentrate on one useful and efficient class of methods, usually called the *quicksort* methods, after the first algorithm of the type.

The method developed originally by Hoare (1961) proceeds as follows. Initially, we choose a value, t, (usually called the *fence)* and split X into two parts consisting of elements that are smaller or larger than t. Thus we have, after the first pass, split X into the form

$$[X_1, X_2]$$

with $X_1 \leqslant t$ and $X_2 \geqslant t$. Next the splitting algorithm is applied recursively to X_1 and to X_2, choosing a new fence each time. Then the split pieces of X_1 and X_2 are split again, and so on. Eventually, provided each fence is within the range of the segment being split, the data will be completely sorted. The operation can occur in place, that is, by writing intermediate results back into the array X.

While this description is admirably simple, it will hardly be sufficient to generate an efficient algorithm. Clearly, a great deal of recursion will take place, and this needs to be done carefully. Recursive computing always involves the stack-type storage discussed in Section **a**. If a procedure, SORT, is to call itself, then it must arrange to put on top of the stack sufficient information to allow the current execution of SORT to resume unchanged when the recursive call is completed. There are several approaches to this, but a simple one is to keep track of all the local variables in SORT, and to write these onto the top of the stack when a recursive call is to occur. On return, the local information is retrieved and the stack height is returned to its previous position. In some form, SORT must always know the current stack position to make this work. (Equivalently, if the stack is internal to SORT we need to know how many levels of recursive calls are currently active.)

Block structured languages (e.g., ALGOL) generally do all this for us automatically. Inevitably this convenience carries some overhead, as data may be put on and retrieved from the stack when not really needed. Just how severe is the penalty depends on the language (particularly, how much control it gives the user over recursion strategy), the particular problem, and the programming style. In the original version of the splitting algorithm, a recursive statement of the procedure is so simple that one expects little unnecessary overhead.

The sorting algorithms based on the ideas in *quicksort* are good general-purpose procedures for internal sorting. The original algorithm was subsequently revised for greater efficiency, notably by Singleton (1969). The algorithm sorts in place, requiring extra storage only for the recursion. Its computational efficiency has been found to be very good.

To be more specific on the last point it is necesary to comment briefly on measurements of efficiency in sorting. Two decisions must be made: how to count costs and what to assume about the initial ordering of the data. Cost measurements may be empirical (processor time, billed cost) or may be derived by counting specific operations (comparisons). Both are needed since empirical measurements apply only to a specific environment and counts are only an approximation to actual cost.

For sorting, a simple approximation is to count only comparisons of pairs of items. Notice that each level of recursion in the *quicksort* algorithms compares each of the N items with (some) fence value. If one could choose the fence to be the median of the data to be split, each piece would be divided in half each time. The total sort then requires essentially $\log_2 N$ levels of recursion, giving a total of $N \log_2 N$ comparisons. This analysis is clearly too optimistic, since computing the median each time would be far too expensive. By estimating the median crudely, however, the expected number of comparisons can be made fairly close to N $\log_2 N$ for most reasonable assumptions about the input data.

The most common approaches to characterizing the input data are either to assume the data are a random sample (i.e., serially uncorrelated) or to find the input ordering among all data with N items that gives the worst case (maximizes whatever cost measure is being used). These two approaches, average performance and worst-case performance, are characteristic of most computational analyses, numerical or nonnumerical. Some attention has also been given to special types of input for sorting, for example, data that is nearly sorted on input. Under the assumption of random input, Singleton's algorithm requires somewhat less than 2N $\log_2 N$ comparisons. Its worst-case behavior requires $O(N^{2})$ comparisons (imagine the fence was always the smallest element of the set), but the required assumptions are implausible in the extreme.

Two extensive studies of sorting costs are given by Loesser (1976; 1974) and Knuth (1973, p. 381). Both give operation counts for several sorting algorithms; the general conclusion from both is that *quicksort* algorithms are competitive with any in-place sorting procedures for general use.

The published algorithm by Singleton (1969), containing both FORTRAN and ALGOL60 programs, is a good general procedure. See Loesser (1976) for specific comparisons and variations. Knuth (1973, 116-119) gives a pseudo-assembly language algorithm version, which might be useful as a model if such low-level coding was desirable.

The advantages of the *quicksort* algorithms are: they sort in place, they are as efficient as any algorithms for moderate-sized data if nothing is assumed about the original ordering or about the set of possible values, and they adapt well to the partial-sorting problem (see the following) and to the sorting of non-numeric data. With respect to the last point, notice that one simply replaces comparisons with an external logical function to compare two items. The disadvantages of the algorithms include: they do not take advantage of data that is nearly sorted already, they do not preserve order of tied items, and better methods exist when there are only a small set of possible values for X[i]. For the first case, bubble sorting is recommended; for the third, radix sorting is appropriate. Either method can be made to preserve order of ties. For these and other special cases, see Knuth (1973) and Lorin (1975).

Partial Sorting. An interesting problem related to sorting and occurring in some practical data analysis techniques arises when only some of the order statistics are required (Chambers, 1971). Suppose X_* stands for the result of sorting X, and let i be a vector of k + 2 integers. For convenience we let i[1] = 0 and i[k + 2] = N + 1. The remaining k elements of i are subscripts in X and we assume they are ordered, so that

$$0 = i[1] < i[2] \cdots < i[k+1] < i[k+2] = N+1$$

Then we say that a permutation X_{**} of X *partially sorts* X with respect to i if the following conditions hold:

1. $X_{**}[j] = X_*[j]$ for $j = i[2],\dots,i[k+1]$;
2. If $i[L] < j < i[L+1]$,

$$X_{**}[i[L]] \leqslant X_{**}[j] \leqslant X_{**}[i[L+1]]$$

To handle the trivial cases when L = 1 or L = k+1, we assume $X_{**}[0] = -\infty$, $X_{**}[N+1] = +\infty$. The first condition says that X_{**} gives us all the order statistics of X corresponding to elements of i. The second says that those order statistics bound the intervening subsets of X_{**}.

Partial sorting has applications to various techniques for *robust estimation* in which estimates are based on some of the order statistics or treat data in different sections of the ordered sample differently. For example, the *midmean* of a sample is defined, roughly, as the average of the central half of the order statistics. Thus, we could partial-sort X with k = 2, i[2] = N/4 and i[3] = 3N/4.

It happens that *quicksort*-style algorithms are splendidly suited to partial sorting. One merely skips all further splitting on any segment that contains none of the subscripts in i. Aside from the extra bookkeeping in manipulating i, such an algorithm must be cheaper than full sorting. Chambers (1971) gives a partial-sorting algorithm based on Singleton's sorting algorithm and shows some experimental results on relative costs. When N is fairly large, say on the order of 100 or more, and k is small compared to N, significant savings occur. A FORTRAN algorithm is included. See also Payne (1973) for a form of external partial sorting.

For much more detailed discussion of sorting, both internal and external, see Knuth (1973) or Lorin (1975). The former gives a lengthy and readable discussion of many mathematical and theoretical questions, as well as an extensive introduction to basic techniques. The latter discusses practical questions in the design, implementation, and use of sorting procedures, particularly from the system programmer's viewpoint.

e. Searching and Table Look-up

We now consider the problem of looking for a given item in a set of items. This is the *table look-up* problem, in the literature of which the items are usually called *keys.* A key may be a character string, integer, or real number, or a set of several such values. In the most elementary situation all we need to do with keys is to establish whether two of them are the same. For efficient search techniques, the keys must have an ordering. For hash-coding procedures, the keys are assumed to define actual numerical values.

Each key will generally identify some additional information, commonly called the *record* associated with the key. Usually the point of the search is to find and/or modify the record associated with the key. The general problem is the following. We are given a *table,* T, of N keys (and the associated records) and a key, x, possibly identical to one of the keys in T. The problem is to *find* an element T[j] that equals x, if any. In the event that x is not in T we may also require a procedure to *insert* x so that a subsequent search of T will find x. Finally, we may need a mechanism for *deleting* x if it is found in T.

In considering the economics of table look-up, an important distinction is whether the entire table can be accessed at the same time. Tables of names in a programming language or names of variables in a data set, for example, should be small enough to keep internally. In contrast, a table of all entries in a large data base will tend to be too large. By analogy with sorting, these two cases may be called *internal* and *external* table look-up. The recommended methods are either *tree-search* or *hash-coding* for internal and a variation of hash-coding for external look-up.

Internal Table Look-up. The exact operation of searching or hash-coding depends on what information is required if x is found, whether x is to be added to the table if not found or deleted if it is found. Suppose the following case. If x is in the table T, we want to return its position or index. If it is not in the table, we wish to enter it.

Tree search algorithms specify the choice (generally

independent of x) of a first element, t_1, in T to compare to x. If $x = t_1$ we are done. Otherwise the algorithm specifies a next element, t_2 or t_3, according to whether $x > t_1$ or $x < t_1$, by whatever ordering procedure is specified for the keys. In the first case t_2 is the *right successor* of t_1; in the second t_3 is the *left successor*. The comparisons continue until a match occurs or until the required successor does not exist, in which case we conclude that x is not in T.

If we assume the table to be an ordered array, the basic search algorithm is shown in Figure 4. Notice that this simple algorithm behaves well if x is found in the table, but requires considerable work to insert x, more work in most cases, than the search itself. When frequent updating is expected and computing efficiency is relevant, more elaborate techniques may be used. Half the insertion work is saved on the average if T is stored as a double-ended array (compare Figure 1). See also Knuth (1973, Section 6.2). For most applications in data analysis, however, retrieval is much more frequent than updating.

I. Choose an initial point, say $i = N/2$. Set $i_{left} = 1, i_{right} = N$.

II. If x equals k[i], x is in T with index i; return.

III. If $x >$ k[i] then:

a. if $i = i_{right}$, x is not in the table. Insert it as the $(i+1)^{st}$ item.

b. else, set $i = \max(i+1, (i+i_{right})/2)$, set $i_{left} = i+1$, and continue from II.

IV. If $x <$ k[i] then:

a. if $i = i_{left}$, x is not in the table. Insert it as the i-th item.

b. else, set $i = (i+i_{left})/2$, set $i_{right} = i-1$, and continue from II.

Figure 4: Tree Search (Binary Search)

Hash-coding techniques begin by embedding the set T of existing keys in a larger table S of size M. Then a function (the hashing function), say h(x), computes an index in S as a function of the key, x. If S[h(x)] = x then we have found the required key. If

S[h(x)] is *empty,* we conclude that x is not in the table. Otherwise a second function of x is computed, and this element of S is examined. The process continues until either x or an empty element of S is encountered, or until we know that every element of S has been examined. In either of the latter two cases we conclude that x is not in T. The most general procedure would imply the existence of M functions, $h_1(x)$, $h_2(x)$,...,$h_M(x)$, representing the 1st, 2nd,...,M^{th} place in S where we look for x. Actually, nearly all hash-coding uses only two functions: a hashing function, h(x), to compute the first *probe* and a function, next(j), to compute the next probe from the current probe or from x. This simplifies the treatment of long or irregular keys, without degrading the performance of the algorithm seriously. A simple algorithm for hash coding is given in Figure 5.

I. Set i = h(x); nprobe = 0.

II. If S[i] = x, x is in the table with index i; return.

III. If S[i] is empty, x is not in the table. Set S[i] = x to enter x and return.

IV. nprobe = nprobe + 1; i = next(i).

V. If nprobe = M we have searched the entire table; the table is probably full and we should take suitable panic action.

VI. Else, repeat from II.

Figure 5: Hash-coding

The operation of the algorithm is reasonably obvious, except for Step V. Here there are two possibilities: either the table is full, which is essentially unrecoverable, or else we have deleted some items in the probe chain. In any case, executing the full M probes would be disastrously expensive. In any reasonably well-running hash table, Step V will never be executed.

It is relatively easy to make a hash-coding algorithm that works. To achieve a correct result, h(x) must be well-defined; that is, it must always give the same first probe for the same or equivalent keys. For example, one should transform x to remove any nonsignificant characters, to resolve upper case versus lower case, etc., before computing h(x). The function, next(i), must be such that *every* entry of S will eventually be reached by the sequence next(i), next(next(i)), $\cdots$, for any initial i. Given these conditions it is easy to show that the hash-coding algorithm terminates with correct results.

The criterion for a good hashing function is that it produces different values for different keys. Generally there are two phases to hashing: first, to reduce the multiword, possibly variable-length key to a numeric value and, second, to hash this value into the range of subscripts in the table. The ideal hashing function would produce a unique value for each input key. This would be difficult even if we knew all the keys a priori; since we usually do not, it is impossible. The worst performance would be to produce the same value for a large number of keys, in which case we would end up doing a sequential comparison of these to retrieve a particular one.

The catch is that we cannot guarantee against an arbitrarily bad assignment of the input so long as the set of keys are not known in advance. It is *possible* that every key will generate the same h(x), so that as many as N comparisons might be required. The problem is somewhat similar to generating pseudorandom numbers to estimate an expected value (Chapter 7); it is difficult or impossible to *guarantee* anything about the results. One can, however, design the computations in both cases so that it is extremely implausible that the results will deviate very far from average behavior.

In fact, one idea would be to choose a "random function" of x as the value of h(x). As a practical technique, this suggests using x_*, a reduction of x to a single value, as a starting value for a pseudorandom number generator. A very common choice is to take a *multiplicative congruential generator.* If x_* is a numerical value derived from x, the first probe in step I of Figure 5 is

$$i = \text{mod}(x_* \times \lambda, M)$$

For this purpose the table size M should usually be a prime number. See Knuth (1973, 508-509) for further considerations. Brent (1973) gives a FORTRAN algorithm for hashing, using one-word keys; see Problem 4 for a reduction from general keys.

Costs. The running time of search or hash-coding algorithms is frequently approximated by the number of keys tested, plus the cost of insertion if required. This is fairly reasonable for simple algorithms such as we have presented. The simple binary search requires as many tests as the depth of x if x is in the tree (depth being the number of nodes from the first along the path to x) and as many as the depth of the empty node where insertion takes place, otherwise. Thus the cost depends on the structure of the tree. The best tree, in the worst-case sense, is one whose paths all terminate after the same number of nodes. This can be done exactly if the number of nodes is $2^k - 1$ for some k. One uses the median as the first node, the median of the first half as the left successor, etc. Such a tree can of course only be constructed if the keys are completely sorted. As new keys are entered the balance will be destroyed. Setting up such a tree will usually be too expensive, except when we are only searching in a fixed table without doing insertions or deletions. On the other hand, it is obvious that the worst possible sort of tree can require N steps to search among N objects rather than $O(\log_2 N)$ which balanced trees require. Considerations of this sort lead two ways: to different analyses which make binary trees look better and to modified tree search algorithms whose worst-case performance is more acceptable.

The most reassuring analysis is the assumption that all possible orderings of the keys as they arrive for insertion are equally probable. In this case the average number of steps to find an item in a tree of N keys is $O(\log_2 N)$. With equiprobable orderings, the average number of comparisons turns out to be only about $1.4 \log_2 N$ (Knuth, 1973, p. 427).

Possible modifications to ordinary tree search amount to rearranging the table of keys when a new key is inserted, in order to prevent very long search times. The question is how much work we are willing to do with each insertion. There are a number of techniques for producing trees that are good in either the average or worst-case sense. A technique known as *balanced tree search* allows all operations (searching, insertion, deletion) to be done in $O(\log_2 N)$. Thus a modest penalty is imposed on search time, compared to a completely sorted binary tree, while the amount of work on insertion is still of the same order of magnitude as our unsorted algorithm above. For further discussion, see Knuth (1973, 451-469).

By comparison with tree searching, hash-coding techniques usually require more space, less time on the average (at least for fairly large tables), and much more time in the worst cases. If the table is large enough that there are many empty positions (say, if $N/M < 2/3$) a good hashing algorithm should keep the average look-up cost down to less than 1.5 times the cost of the initial probe. The reduction of multi-word keys in a reasonable way is important (see Problem 4). Since look-up is generally much more common than insertion or deletion, algorithms, such as Brent (1973), may reasonably do extra work on insertion to speed up later search for the same key. The most comprehensive single source for algorithms on searching is currently Knuth (1973). Algorithms are described there verbally and in the machine language of a fictitious computer. While both forms require further work before they can be used, the search algorithms are relatively simple to program.

Search and look-up procedures have many and diverse applications, particularly to basic computing software such as compilers and operating systems and to commercial and accounting procedures. When used in data analysis for searching in small tables of names treatment as internal look-up is appropriate. The most important applications, however, are often to the management of large data bases. In this case, some additional considerations apply.

External Table Look-up. We assume now that the table is not immediately accessible. Typically, data must be read from disk or tape, in order to look at an entry in the table. In this case, the cost of the access to the disk or tape may dwarf the other costs. An algorithm that keeps the number of accesses per look-up low will be the most efficient.

Hash coding can be adapted as follows to achieve this result. To each possible value of the hashing function, one associates a block of external storage containing the entries for *all* items producing this value of h(x), with exception of overflow, to be discussed. This block is usually called a *bucket* in data-base terminology. It should be a convenient size to be read at one access, large enough to hold the number of keys likely to hash to it, but small enough to fit easily into memory. Sizes of a few hundred to a few thousand characters are typical. On many machines, the bucket will be a multiple of some intrinsic block size on disk.

Finding a key then involves one access to get the bucket, followed by a search for the key within the bucket. By assumption, optimizing the search is relatively unimportant; often all the keys are just kept in a linked list (Section **a**), which is searched sequentially.

The exception mentioned is that the bucket may become full. Some overflow mechanism is needed, that is, a next(i) function. One may re-hash, or look at the next sequential bucket. The overflow strategy should not be very important. If bucket overflow occurs often, the basic scheme is in trouble. Either the table is too full (the total space allowed for buckets was too small), or the hashing algorithm has done a bad job. These cases are distinguished by overflow in many or in a few buckets, respectively.

With a high-quality hashing function and a reasonably generous space for buckets, the average number of accesses to find a key can be kept to within a few percent of 1. See Knuth (1973, 534-538) for some further discussion.

A further useful trick is to keep in memory a list of all the table information for, say, the last k items accessed, which is searched each time *before* hashing. For applications accessing the same data repeatedly (an interactive computing system, for example) this can further reduce the number of accesses substantially.

f. Using Data Base Systems for Data Analysis.

In selecting or developing data base software for scientific data analysis it is well to consider the characteristics of that application, as these differ in some important ways from other uses of such programs.

The typical applications of current DBMS share certain characteristics.

1. The data is structured in a well-defined way, usually by a hierarchical or network scheme.
2. The integrity of individual data items is very important and the correctness of any operations performed and reports generated should be (one hopes) meticulously verified. Partly as a corollary, the operations performed are frequently specific and standardized.
3. The structure of the data base tends to reach very deeply, so that the amount of data directly recorded in an individual node is usually a small fraction of the total data.

Generally, these systems are oriented towards static structure, highly disaggregated data, and the processing of transactions and/or report generation.

From the characteristics of the applications follow several characteristics of the systems themselves:

1. Most contain a declaration phase, called a *data description language* (DDL) in CODASYL, giving a permanent description of the hierarchical interconnections in the data and, usually, of the type and size of the fields in which actual data is stored.
2. Manipulation of the data base tends to be relatively high-level and "packaged," particularly as many systems hope to involve administrators and other nontechnical personnel in use of the programs.
3. The general computational design of the systems does not usually stress the internal modularity that would invite nonstandard use of pieces of the data structure.

Scientific data analysis also depends on the availability of data management, whether or not this dependence is explicitly recognized. When the data is to be analyzed on a computer there must be some software to provide the management. The purpose is to facilitate analysis of the data and hence better understanding of the underlying processes. Such data has both strong similarities to and sharp differences from the more typical DBMS application.

Frequently the data exhibits important hierarchical or other interconnections, which make the use of DBMS programs attractive. At the same time, the nature of the connections will not be precisely or rigidly defined, and one will want to adapt and vary the structure imposed to suit the analysis. Since analysis is a major purpose of the data base, it is important that the programs be able to interact freely with libraries and algorithms for data analysis, and that the DBMS software itself not be too burdensome in size or time requirements. Also, the DBMS software is likely to be used by scientists or professional programmers rather than administrative or business personnel. The nature of much scientific data is that it is frequently large in volume and contains regular subsets of data (for example, vectors or two-way arrays of observations), which, however, may vary in extents. Thus a combination of great flexibility of definition with efficiency of access is important.

In choosing or developing data base software, flexibility, efficiency, and ease of use must be balanced against one another. For data consisting of large subsets that are regular or nearly regular, very considerable savings in computing costs may occur if the data structure is treated as an array, rather than by hierarchical relationships. During the analysis, the access to portions of the data will tend to be in terms of linear sections of the array, as defined in

Equation (3) in Section **b**. Given the starting address of the array in external storage and the extents of the array, the address in *each* item is then known, without further access to the external medium. If, instead, the array A were broken up in hierarchical fashion, the retrieval algorithm must first determine the location of each node before accessing the data in that node. This table look-up or search may itself require access to external storage (e.g., if a large table is used), and the consequent cost of access may be much increased. As noted at the end of Section **b**, the efficiency of access to linear sections of an array depends further on the spacing of the elements. If an array is to be accessed one row at a time, one would like to store it by rows, to reduce the number of accesses required.

For many applications, much of the efficiency noted may be retained while still benefiting from the auxiliary information and flexibility of data structures. Consider the relatively simple structures in Section **c**, above. Notice that series and dmbyvar both consist of a combination of character string descriptive information and data. The individual pieces such as *values* in (5) are arrays and thus could be efficiently accessed, provided one could get at their starting address. The best way to manage this will depend on the details of implementation. The declarations in (4) to (7) imply that the data has already been made available, but suppose we interpret a pointer or **ref** as an address on a disk or other external medium. Then if we rewrite (7) as

mode dmbyvar = **struct**(
[1:8] **char** name, (9)
ref [,] **char** unitname,
[1:10] **ref** series variates)

the structure so defined provides us with a vector of ten pointers to find the data in each variate.

Ideally, one would like the data analysis and data base management to be fully compatible, so that one could potentially access data from the data base and immediately apply statistical algorithms to it. Two factors currently make this less practical. Many data base management systems make it difficult to invoke a FORTRAN algorithm, for example, on data in the system. At the same time, most statistical algorithms require their input to be regular vectors, matrices, or arrays and cannot treat hierarchical or other irregular structures.

Most statistical users of DBMS, therefore, will need some facility for expanding portions of the data base into regular form and copying it to some external dataset. This may have the additional advantage of making repeated access to the data more efficient. However, *updating* the data presents problems. Corrections and changes made in either copy will not usually be reflected in the other. Depending on the reason for the change, we may not want them to be (particularly in the case of editing of the statistical copy for purposes of analysis). If updating is to be done on both, some link, either by a DBMS operation or manually, must be made between the two copies. Confusion is usually kept down, however, by treating the statistical copy of the data as scratch storage, and regarding only the data base copy as permanent. These issues are often discussed in the DBMS literature as the *copy problem.*

g. Summary and Recommendations

Of the topics discussed in this chapter, storage management, sorting, and searching are likely to generate the need for specific algorithms. The most common requirement for storage is scratch space on a stack for temporary use in other algorithms. This form is provided in some languages and may be simulated explicitly in any case, as in the FORTRAN system in Fox, Hall and Schryer (1975). More complex problems, such as data-base management, may require storage allocated on demand. One approach to this storage is through list-style structures, as in Bray (1974).

Sorting problems in which the data can be held internally may be handled efficiently by the *quicksort* type of algorithm, among others. A careful implementation of this method is given by Singleton (1969) and an adaptation to partial sorting by Chambers (1971). Lorin (1975) reprints these algorithms and most others published up to the mid-1970's. External sorting should best be handled by an efficient sort system, such as are provided by many computer manufacturers. Detailed discussions are found in Lorin (1975) or Knuth (1973).

The major techniques for searching or table look-up are binary tree searches based on comparisons and hash-table techniques, based on the computation of successive probes. The former are generally simpler and give better guaranteed performance. Hash coding methods give better average performance, assuming uniformly random keys, and are preferred for large scale applications such as data-base management. For these applications, the table should be divided into *buckets* corresponding to the hashed values, so that only one external access is usually required. Algorithms for many search procedures are found in Knuth (1973), although not in a standard programming language. Brent (1973) gives a good hashing algorithm in FORTRAN.

The discussion of arrays in Section **b** and data structures in Sections **c** and **f** concentrated on ways to obtain both flexibility and reasonable efficiency. Taking advantage of the storage scheme for arrays, and particularly of the concept of a *section* of equally-spaced elements, leads to simple implementations of efficient, general operations. For data-base software, the ability to combine hierarchical structure with efficient treatment of regular substructures as arrays is very desirable.

Problems.

1. General k-way arrays, where k may be specified during execution, are useful in a number of statistical analyses (such as analysis of variance and contingency table analysis). With the exception of APL, most programming languages require array declarations to have a fixed number of subscripts. Design a set of algorithms to simulate the general structure in the language of your choice. If the language allows, declare the array as a data structure.

2. A k-way *table with margins* is defined from a k-way array by allocating an extra level, m_j+1, for each subscript in A. Margins are computed by applying some operation (means, totals, medians, or whatever) to a set of m_j elements which agree in all subscripts except the jth, and storing the result in the (m_j+1) element. Show how to generate all the margins in a table, using the idea of array sections and treating the table as a one-way array of $\prod(m_j+1)$ elements.

3. (Bubble sort) Design a sorting algorithm that works as follows. Compare X[i] and X[i+1] for i=1,...,N′, and interchange them if X[i] > X[i+1]. If an interchange occurs for any i, set N′ to the largest such value of i and repeat. Show that the algorithm sorts X if N′=N initially. If possible, implement this algorithm and compare it in size and speed with a *quicksort* algorithm. See also Knuth (1973, 106-111).

4. Suppose the keys in a table look-up are character strings of arbitrary length. Design an algorithm to reduce the keys to single integers. The algorithm should give distinct results for typical input strings, such as alphanumeric strings up to, say, two words long. Getting a good solution will probably require considering the key as a *bit string* and using operations such as rotation and exclusive-or (see section **4.a**).

CHAPTER FOUR

Numerical Computations

a. The Representation of Numbers; Bit-strings.

Throughout most scientific computation we can ignore the code used to represent integers or floating-point numbers in the computer. A small set of values giving the permissible range and/or accuracy are sufficient except for studying some intractible bugs and for particularly nonstandard or delicate calculations. Occasionally, however, it is important to understand the various representations in use. In this section we consider commonly used representations.

Historically, number systems were introduced in the order: natural (unsigned integer), signed integer, rational, real, complex. The same order makes sense in discussing computer representation of numbers, except that rational numbers play a relatively slight part. As we proceed, the possibilities for divergence in representation increase.

Most contemporary computers use binary (i.e., two-state) elements to code numeric data. Unsigned integers are then represented as strings of binary digits:

$$i = \sum_{j=1}^{L} b_j 2^{L-j}$$

where b_j is the jth *most significant bit* in the representation of i. The internal storage of i can be represented schematically as

$$b_1 \quad b_2 \quad \cdots \quad b_L \tag{1}$$

The number, L, of binary elements required to code i is called the *length* of the representation. A generally harmless ambiguity is to refer to the jth bit to mean either the jth binary element or its numerical value. The range of unsigned integers of length L is $0 \leqslant i \leqslant 2^L - 1$.

When the values of specific bits are relevant, data is generally printed using a base of the form $\beta = 2^k$. Then,

$$i = \sum_{m=1}^{\lambda} h_m \beta^{\lambda - m} , \tag{2}$$

and the value h_m depends on k bits; namely,

$$h_m = \sum_{i=1}^{k} b_{k(m-1)+i} 2^{k-i}. \tag{3}$$

The point of using 2^k as a base is that the specific value of any bit can be found easily from looking at a single h_m. In order to facilitate arithmetic, base 2^k, and to display i in a readable way, one chooses k so that 2^k is near 10. Specifically, the values $k = 3$ or $k = 4$ are nearly universal. Unsigned integers base 8 and 16 are usually called octal and hexadecimal (hex to most programmers).

The consideration of *signed* integers produces several possible representations. The commonly used forms all add one bit, the *sign bit,* with the value 0 for positive numbers and 1 for negative. There are at least three methods, however, commonly used to represent the magnitude of a negative number. Figure 1 defines the three systems. In each case, we show the *unsigned* $(L + 1)$-bit integer which is used internally to represent a *signed* integer in $(L + 1)$ bits. Note that the formulas cannot always be applied in a machine with $(L + 1)$-bit arithmetic to convert from one representation to another; for one thing, the range column of the figure shows that the range of numbers is not exactly the same. The range and the uniqueness of zero are the major practical points to be kept in mind.

	$-(i)$ as an $(L+1)-$bit Number	Range	Representations of Zero
Signed–Magnitutde	2^L+i	$[-(2^L-1), 2^L-1]$	$0, 2^L$
2's Complement	$2^{L+1}-i$	$[-2^L, 2^L-1]$	0
1's Complement	$2^{L+1}-i-1$	$[-2^L-1, 2^L-1]$	$0, 2^{L+1}-1$

Figure 1: Representation of Integers

A fourth representation should be mentioned, which is sometimes applied in floating-point systems; namely the *excess* 2^L system. We represent the signed L-bit integer, i, by the unsigned $(L + 1)$-bit integer $i + 2^L$.

Rational numbers, as noted, play a relatively small part in data analysis. A rational number could be represented as a pair of integers, that is, the rational number $r = i/j$ is represented by the pair, (i,j). Only one of the integers need be signed, of course, and the representation is not unique. Arithmetic can be defined on such numbers, and many significant problems handled directly in terms of rational arithmetic, notably in combinatorial mathematics, symbol manipulation and number theory. We do not pursue rational-number calculations here; the interested reader is referred to Knuth (1969, Section 4.5) for an introduction.

In contrast to rationals, *fractions* are crucial to numerical computing. If we interpret an L-bit integer, i, as the fraction, $i/2^L$, then we can use L bits to represent positive fractions; namely, the bits $b_1,b_2,\ldots,b_L$ in (1) represent

$$f = \sum_{j=1}^{L} b_j 2^{-j}. \tag{4}$$

By adding a sign bit and choosing a representation for signed integers, signed fractions may be represented. Be warned that some computers use *different* representations for signed fractions and signed integers, e.g., two's complement for integers, signed-magnitude for fractions.

By far the majority of scientific data analysis involves *real* or *floating-point* numbers. A floating-point number can be represented as the product of a fraction and an exponent base raised to a signed-integer power; namely,

$$x = f \times \beta^e. \tag{5}$$

Generally f is called the *fraction* or *mantissa* and e the *characteristic* or *exponent.* The number is then stored by coding, in some form, the pair (e,f). Since the fraction is represented as an integer, this amounts to representing x as a pair of signed integers. In some machines, the physical representation is exactly this; in others, the sign bit for the fraction is stored separately. Two common representations are

$$e \quad f \tag{6}$$

and

$$s \quad e \quad f_+ \tag{7}$$

where s is the sign bit of f; i.e., $f = (s,f_+)$. The exact set of floating-point numbers representable and the rules for operating upon them depend on the choice of representation for e and f. These may be different, as noted. The current IBM 370 uses excess - 2^6 for the exponent, signed-magnitude for the fraction, and two's complement for ordinary integers. In contrast the Honeywell 6000 uses two's complement for all three choices.

Floating-point operations on nearly all contemporary machines work with *normalized* floating point numbers, for which the fraction is always shifted left until its most significant digit is non-zero. For binary machines, this means that the absolute value of f will always be at least 1/2. The precise implications for the bit-patterns depend again on the choice of representation, as does the exact range. For signed-magnitude, the first bit of f will always be 1 and the range $-1/2 \leqslant f \leqslant 1/2$. For two's complement the first bit is 1 for positive f and 0 for negative f, and $-1/2 < f \leqslant 1/2$.

There remain three more parameters of the floating-point representation: the lengths of the fraction and exponent, and the choice of a base, β. For a chosen total number of bits to represent x and chosen base, an increased length for e gives a greater range and increased length for f gives greater accuracy, i.e., a closer approximation to a chosen real number in the range.

The exponent base, β, will nearly always be a power of 2 for binary machines, since only then will the full range of possible exponents be used. The choice of power is more subtle. It was shown by Brown and Richman (1969) that for a fixed total number of bits, $\beta = 2$ could give, with some choice of the lengths of e and f, at least as good accuracy and larger range as any choice, $\beta = 2^i$ for $i > 1$. The definition of accuracy used was of the minimax form, including such criteria as the maximum relative difference between a given real number and its closest floating-point representation.

Most machines and most programming languages make some provision for more than one floating-point representation. By using a longer string of bits than occupied by standard floating-point numbers, a representation with greater accuracy and/or greater range is obtained. We see in the next section that such *extended precision* numbers are useful in certain numerical calculations. By using extended precision for a few selected calculations, overall guarantees of accuracy can be much improved without significant increase in computing cost. Generally, computer hardware provides only a few options on floating-point precision (usually two, sometimes only one). More extended precisions can be simulated by programs for the basic numeric operations.

The most common situation is that extended-precision numbers occupy twice the number of bits used for ordinary numbers. Notice that in this case, it is possible to more than double the number of bits for the fraction, even if we also wish to double the length of the exponent as well. Many of the important results of floating-point error analysis use this assumption. To make the distinction clear, we shall use the term *double precision* to imply the following condition: if t is the length (excluding sign) of the fraction in an ordinary floating-point number, the length of the fraction in a double precision floating-point number is at least $2t+1$. Programming languages differ in the provision for extended precision. FORTRAN includes the double precision declaration, but does not specify its meaning. PL-1 allows precision (i.e., the length of f) specified explicitly up to a machine-dependent limit. ALGOL68 intends the long mode to imply double precision but does not appear (as far as the author can tell) to require the above condition. A few languages, such as APL, provide no extended precision facilities.

Bit-strings. A different set of operations arise when the number in (1) is regarded strictly as a vector whose elements b_j are 0 or 1. Bit-strings arise naturally in some data management procedures (see Problem 4 of Chapter 3), in symbolic calculations and in some special techniques, such as the analysis of variance and random number generation.

Operations which take one or two such vectors and return a similar result are:

$$c=\mathrm{AND}(b,b')$$
$$c=\mathrm{OR}(b,b')$$
$$c=\mathrm{XOR}(b,b')$$
$$c=\mathrm{NOT}(b)$$

In the first three a bit of c is 1 if both, at least one and exactly one, respectively, of the corresponding bits in b and b′ are 1. The NOT operation changes 0 to 1 and conversely. To operate effectively on bit-strings, one also needs to be able to *shift* and *rotate* bit-strings. In shifting, for example, left by k bits, b_j is moved to the position $(j - k)$. The leftmost k bits are lost, and the rightmost k bits filled with zeroes. In contrast, rotating left by k bits moves the leftmost bit into the kth rightmost bit, and so forth. The operations might be written:

```
c=SHIFT(b,k)
c=ROTATE(b,k)
```

with a convention that motion is left if $k > 0$ and right otherwise.

Some programming languages provide at least a portion of these facilities. Otherwise, some set of such operations need to be implemented outside the language (Problem 1). A distinction must be made between bit-strings of arbitrary length and single words. Machine language operations often provide the latter, but the more powerful general operations require more programming.

b. Floating-Point Operations; Error Analysis

An essential characteristic of floating-point numbers is that they are not in exact correspondence to the real numbers they represent. This can be stated two ways. A given real number, x_*, must be *rounded to,* i.e., approximated by, a representable floating point number, x. Conversely, a given floating-point number x can be taken to represent an *interval,* I, in the set of real numbers such that each x_* in I would be rounded to x.

This nature of floating-point numbers makes it desirable to control, bound, or estimate the *rounding error* of a floating-point algorithm; that is, the difference between the effect of the algorithm when performed with machine operations on floating-point numbers and the effect of the corresponding operations on true real numbers (which for most purposes can be thought of as floating-point numbers with an infinite number of bits for e and f).

By far the most influential technique for studying rounding error is *floating-point error analysis.* More precisely, this term usually refers to *backwards floating-point error bounds,* either relative or absolute. Given an algorithm A operating on inputs $x_1, x_2, \cdots, x_m$ to produce outputs $y_1, \cdots, y_n$, one wishes to prove, under reasonable assumptions that the computed outputs would have been produced by *exact* calculation with inputs $x_1^*, x_2^*, \cdots, x_m^*$ and to show that differences,relative or absolute, between x_i^* and x_i are bounded. When the bounds are attractively small, one has guaranteed, in this sense, the accuracy of the algorithm.

Equations (9) to (12) below summarize the basic error analysis for floating-point operations. These bounds are the building blocks for the error analysis of larger algorithms such as the linear calculations of Chapter 5. Before beginning the detailed discussion, we should re-emphasize the purpose of the error analysis, since this may be misinterpreted. Based on some reasonable assumptions about machine characteristics, bounds are derived on errors relative to exact arithmetic using the floating-point numbers *stored in the computer.* The methods of proof and the results obtained are not necessarily appropriate or valid for the study of observational errors or uncertainty in the original data. This is particularly the case when such errors may be large or their magnitude difficult to assess. Unfortunately, there exist many confusing statements on the topic, even on the dust-jacket of the pioneering book by Wilkinson (1963). A particularly unfortunate tendency is to assume that computational accuracy implies some corresponding statistical validity to the results. See Section **5.f** for the case of linear regression.

The four basic floating-point operations are add, subtract, multiply and divide. Generally, the result of any of these operations is stored in a machine register. For many modern machines this register has greater precision than the ordinary floating-point numbers. The following error bounds assume that the intermediate result is held in a *double precision* register, using the definition in Section **a**.

Let $x_i = f_i\beta^{e_i}$ for $i = 1,2$ be floating-point numbers as in (5). Then one can show that the basic operations produce the best possible answer, in the sense that the intermediate results are correct to more than t digits, and therefore, the only inaccuracies are in the rounding of the result. Let us prove this for addition, to illustrate the argument.

First, suppose that f_1, f_2 are taken as real fractions terminating after t bits; i.e., we consider the real numbers x_i with exactly the value of $f_i\beta^{e_i}$. Then the *exact* sum satisfies the equation

$$x_i + x_2 = (f_1 + f_2\beta^{e_2-e_1})\beta^{e_1}, \tag{8}$$

where for convenience we assume $e_1 \geqslant e_2$. We argue that (8), in fact, describes an algorithm to produce the desired result. The fraction f_2 is loaded into a double-precision register and shifted $i\times(e_1-e_2)$ places to the right, where $\beta = 2^i$. This result should then be rounded to double-precision accuracy, and f_1 added to it. (On some machines the shifted fraction is truncated, rather than rounding to the nearest bit.) On adding, the sum may be greater than 1 but is less than 2, so that at most one right shift is required. The maximum cancellation is t places (for example, on adding 1 and $-(1 + 2^{-(t-1)})$). In the case of cancellation, notice that there is generally *no* rounding error, because all the bits in the sum of the fractions will be retained. This seems paradoxical unless we keep firmly in mind that rounding error refers to the difference between the computed and exact sum of the *floating point* numbers. If x_i are only approximations to the "real" numbers, cancellation can produce an answer with high relative error compared to this "real" sum. Here is the first and simplest example of the important distinction between rounding error and *sensitivity*. Although adding with cancellation produces no rounding error, it may be very sensitive to perturbations of the x_i and in this sense is an ill-conditioned operation.

Insofar as rounding error is concerned, the description shows that, in any case, the answer in the register is in error by at most one in the last place, i.e., 2^{-d+1}. Rounding the result for storage may produce a relative error of 2^{-t} in the single precision value.

There is also the possibility that *overflow* or *underflow* will occur if the exponent of the normalized result does not lie in the permitted range. Overflow nearly always destroys the numerical validity of the results. Underflow may do so also, although since most computers give zero as the result, the consequences may not be quite so severe. Nevertheless, high-quality numerical algorithms should protect against both overflow and underflow whenever possible. Most existing algorithms do not, but there has been increasing interest in the possibility. See Problem 2 for an example, and the ROSEPACK system in the Appendix.

Assuming the sum is representable, we can describe the error in addition as follows: the computed sum of x_1 and x_2, say $FADD(x_1,x_2)$ satisfies

$$FADD(x_1,x_2) = (x_1 + x_2)(1 + e) \tag{9}$$

where $|e| \leqslant 2^{-2t}$ while the result is in the register, and $|e| \leqslant 2^{-t}$ when the value is stored in a single precision data item. Subtraction gives identical results.

For the discussion of the next section, it is worth noting that we only need to round when we store the result, not while the values are in the intermediate register. The arguments for multiplication and division are similar, but simpler. For multiplication, the *exact* product can always be held in the register, so that the computed product $FMPY(x_1,x_2)$ satisfies

$$FMPY(x_1,x_2) = x_i \times x_2(1 + e) \tag{10}$$

with $e = 0$ when the result is in the register and $|e| \leqslant 2^{-t}$ when stored. For division, there may not be as attractive a result. Many machines with double precision registers still generally produce only a t-bit dividend when dividing f_1 by f_2 (hopefully this is at least correctly rounded). Then

$$\text{FDIV}(x_1,x_2) = (x_1/x_2)(1 + e) \tag{11}$$

where $|e| \leqslant 2^{-t}$ either in the register or stored. Equations (9) to (11) form the basis for error bound estimation in floating-point calculations.

Most medium-to-large modern computers provide double-precision registers for floating-point arithmetic. A machine that does *not* should not be used for any delicate numerical calculations unless it had a very high single precision. The error analysis with single-precision registers is useful, however, for two reasons. First, it will be used in developing some bounds for more extensive calculations. Second, programmers who follow the practice of doing *all* calculations in double precision must use the latter analysis unless their machine has quadruple-precision registers, or the equivalent. This latter point is unfortunately often overlooked in comparing the error for single- and double-precision methods.

The error bounds for basic arithmetic are as follows:

$$\begin{aligned} \text{FADD}(x_1,x_2) &= x_1(1 + e_1) + x_2(1 + e_2) \\ \text{FMPY}(x_1,x_2) &= x_1x_2(1 + e) \\ \text{FDIV}(x_1,x_2) &= (x_1/x_2)(1 + e) \end{aligned} \tag{12}$$

where $|e| \leqslant (1.5)2^{-t}$ and $|e| \leqslant (2)2^{-t}$, whether we are concerned with registers or with stored data. For derivations, see Wilkinson (1963, 11-13).

The book by Sterbenz (1974) gives a detailed discussion of most of the topics of this and the following section. The treatment is somewhat more current and exhaustive (at least with respect to hardware and programming languages) than Wilkinson (1963) and includes a number of special topics, like statistical error analysis and algebraic formulations of error analysis. It omits, however, some critical error results such as inner-product accumulation.

c. Operations on Arrays

Throughout Chapter 5, we consider numerical operations on one-way arrays (vectors) and two-way arrays (matrices). Such operations are sufficiently basic to scientific computing to have had a significant influence on the design of programming languages (as in Chapter 3) and even on hardware design. In this discussion, vectors are dennoted by lower-case letters (u, v, · · ·), matrices by upper-case letters (X, Y, · · ·), and square brackets denote elements (u[i], X[i,j], · · ·).

The most important basic operation on vectors is the *inner* or *dot product,* which we write u·v assuming u and v both have, say, n elements. The error analysis of this operation is of critical importance, both because it underlies so much other work, and because the error bound obtainable is so small that quite comfortable guarantees of accuracy emerge. We now describe this operation; namely,

$$u \cdot v = \sum_{i=1}^{n} u[i] \times v[i]. \tag{13}$$

The following steps are a typical implementation of inner product. We assume that RA and RB are double-precision locations, either two registers or one register and one storage location. We then can do double-precision accumulation, meaning that all intermediate results are in double precision.

I. Set RA = FMPY(u[1],v[1])

II. For i = 2 to n:

(a) Set RB = FMPY(u[i],v[i])

(b) Add RB to RA in double precision (where RA is a double precision variable this may mean add RA to RB and return to RA).

The assumption of at least one double-precision register makes this little more expensive than single-precision accumulation. Only the addition in Step II(b) is actual double-precision arithmetic.

As noted in Section **b**, the product is held *exactly* in RB so that only the errors in the addition count. These are covered by the single-precision result (12) but with t replaced by the number of bits in the double-precision registers, say d. When i = 2 in step II(b), we have

$$RA = u_1v_1(1 + e_1) + u_2v_2(1 + e_2)$$

with $|e_i| \leqslant (1.5)2^{-d}$. After the next step

$$RA = (u_1v_1(1 + e_1) + u_2v_2(1 + e_2))(1+e_3) + u_3v_3(1 + e_4)$$

and so on. Upon completion of the algorithm, but while the result is still in RA, we obtain the result (using a rather special index on the errors):

$$\begin{aligned}RA = (...(u_1v_1(1 + e_{1,1}) &+ u_2v_2(1 + e_{1,2}))(1 + e_{2,1})\\ &+ u_3v_3(1 + e_{2,2}))(1 + e_{3,1})\\ &+ u_4v_e(1 + e_{3,2}))(1 + e_{4,1})\\ &\quad\cdot\\ &\quad\cdot\\ &\quad\cdot\end{aligned}$$

$$= u_1v_1(1 + e_{1,1}) \prod_{j=2}^{n} (1 + e_{j,1})$$

$$+ \sum_{i=2}^{n} u_iv_i(1 + e_{i-1,2}) \prod_{j=i}^{n} (1 + e_{j,1}). \qquad (14)$$

The double index, $e_{j,1}$, $e_{j,2}$, for the first and second terms in the jth addition makes the number of error terms clear. Each $e_{j,k}$ satisfies $|e_{j,k}| \leqslant (1.5)2^{-d}$. Finally, the returned value will be $(1 + e)RA$ with $|e| \leqslant 2^{-t}$. To simplify notation, define $e_{0,2} = 0$. Then

$$DOT(u,v) = \left[\sum_{i=1}^{n} u_iv_i(1 + e_{i-1,2}) \prod_{j=i}^{n} (1 + e_{j,1})\right](1 + e) \qquad (15)$$

where $e_{0,2} = 0$, $|e_{j,k}| \leqslant (1.5)2^{-d}$ otherwise and $|e| \leqslant 2^{-t}$. On the assumption that $d > 2t$, we can write the above as

$$DOT(u,v) = \left[\sum_i u_iv_i(1 + e_i)\right](1 + e) \qquad (16)$$

with, now

$$\begin{gathered}(1 - (1.5)2^{-2t})^n \leqslant 1 + e_1 \leqslant (1 + (1.5)2^{-2t})^n\\ (1 - (1.5)2^{-2t})^{n-i+2} \leqslant 1 + e_i \leqslant (1 + (1.5)2^{-2t})^{n-i+2}\\ |e| \leqslant 2^{-t}\end{gathered} \qquad (17)$$

for $i \geqslant 2$. These bounds apply also to the *normalized* inner product $u{\cdot}v/c$ for some chosen scalar c

$$DOTN(u,v,c) = \left[(\sum_i u_iv_i(1 + e_i))/c\right](1 + e) \qquad (18)$$

with e_i, e satisfying (17).

Note that, while DOT is in the double-precision register, $e = 0$. However, the same is not generally true of DOTN. In order to retain double-precision accuracy, a double-precision divide must be used when scaling by c.

While (17) are the most direct bounds resulting from error analysis of inner products, simplifications based on assumptions about n and t are sometimes useful. If n is small compared to 2^t, then

$$|e_i| \leqslant 1.5(n-i+2)2^{-2t+\alpha}, \tag{19}$$

where α is small, but does depend a little on n. Specifically, if there is an a << 1 such that

$$\begin{aligned}(1 + (1.5)2^{-2t})^n &< 1 + (1.5)n(1+a)2^{-2t}\\ (1 - (1.5)2^{-2t})^n &> 1 - (1.5)n(1+a)2^{-2t}\end{aligned} \tag{20}$$

then we can take

$$\alpha = \log_2(1+a). \tag{21}$$

For applications to matrix arithmetic, it is useful to have a formula directly involving $u \cdot v$. The use of (20) with a slightly larger α is sufficient to combine $(1 + e_i)(1 + e)$ into just $(1 + e_i)$ in (16) with $\dot{e}_i$ still bounded by $2^{-2t+\alpha}$ as in (19). This gives

$$\mathrm{DOT}(u,v) = u \cdot v(1 + e) + \sum_i u_i v_i e_i \tag{22}$$

and the similar form for normalized inner product:

$$\mathrm{DOTN}(u,v) = (u \cdot v/c)(1 + e) + \sum_i (u_i v_i/c)e_i. \tag{23}$$

These are the most frequently used bounds on error in inner products. They say that, for the range of n and t considered, the *absolute* error in computing $u \cdot v$ on $u \cdot v/c$ is $O(2^{-t})$ independent of n unless there is so much cancellation (terms $u_i v_i$ large and opposite in sign) that the e_i terms are inflated by $O(2^t)$. The *relative* error, however, can be large either in this case or when $u \cdot v$ is nearly zero and some cancellation occurs. These properties are at the heart of numerical linear algebra; designing accurate algorithms in the light of (22) and (23) is the key.

Turning to operations on arrays returning other arrays as results, we find two types of error bounds useful. The simplest are bounds on the error in individual elements of the result, which follow easily from analogous scalar operations. Sometimes one wants a summarizing bound for the entire array, however, and this requires a *norm;* that is, a measure of the size of the vector or matrix. There are numerous such norms in use, and mathematical studies of linear algebra are often based on axiomatic definition of an arbitrary norm.

For our numerical purposes in Chapter 5, we use almost exclusively the Euclidean or L_2 vector norm, $||u||$, defined by:

$$||u||^2 = \sum_i u[i]^2; \tag{24}$$

i.e., the square root of the sum of squares. Matrix norms can be defined either directly or (in several ways) from a choice of vector norm. We shall most frequently use the direct analog to (24), the Euclidean norm defined by:

$$||Y||^2 = \sum_i \sum_j Y[i,j]^2. \tag{25}$$

The important matrix operations are scalar multiplication, matrix addition, and matrix multiplication, written here as cY, $Y_1 + Y_2$, $Y_1 \cdot Y_2$, respectively. For these we can derive error bounds as follows. Define E_S and E_A by

$$\text{SMPY}(c,Y) = cY(\mathbf{1} + E_S) \tag{26}$$

$$\text{MADD}(Y_1,Y_2) = (Y_1+Y_2)(\mathbf{1} + E_A) \tag{27}$$

where the left sides represent the computed arrays returned in single-precision and **1** is a matrix with 1.0 in all elements.

From simple extensions of the results of Section **b**, we can show that

$$|E_S[i,j]| \leqslant 2^{-t}$$

$$|E_A[i,j]| \leqslant 2^{-t}.$$

The case of matrix multiplication is not quite so easy. Suppose the computed product goes to a matrix S; i.e.,

$$S = \text{MMPY}(Y_1,Y_2).$$

Then from the analysis of the dot product:

$$\begin{aligned} S[i,j] &= \text{DOT}(Y_1[i,],Y_2[,j]) \\ &= Y_1[i,]\cdot Y_2[,j](1 + e^*_{ij}) \\ &\quad + \sum_k Y_1[i,k]Y_2[k,j]e^*_{ijk} \end{aligned} \tag{28}$$

where we denote by $Y[i,]$ and $Y[,j]$ the ith row and j^{th} column of Y. Under the assumptions leading to (22),

$$|e^*_{ij1}| \leqslant 1.5n2^{-2t+\alpha}$$

$$|e^*_{ijk}| \leqslant 1.5(n-k+2)2^{-2t+\alpha} \quad k \geqslant 2 \tag{29}$$

$$|e^*_{ij}| \leqslant 2^{-t} .$$

This is the basic error bound on elements in matrix multiplication. It can be translated to refer to norms, giving

$$||S - Y_1 \cdot Y_2|| \leqslant 2^{-t}||Y_1 \cdot Y_2|| + 1.5n2^{-2t+\alpha}||Y_1|| \; ||Y_2||. \tag{30}$$

This follows rather easily from (28).

Algorithms in numerical linear algebra are usually based on the related ideas of transformation and decomposition. The idea of transformation is to apply to a general matrix a sequence of linear transformations producing a matrix of a special form. The corresponding idea of decomposition is to express the general matrix as the *product* of several matrices of special form.

The solution of a numerical problem is then carried out in three stages. First, the data is transformed. Second, the solution to the problem is computed for the transformed data. (Presumably, this is easier.) Third, the answer is transformed back to the original form, if necessary.

Chapter 5 deals with computations for linear models in greater detail. For a general introduction to the field, the text by Stewart (1973) is recommended, although the treatment is not particularly directed toward data analysis.

d. Approximation by Rational Functions

So far all the numerical operations discussed have been those typically found in the hardware of a computer: add, subtract, multiply, and divide. Numerical calculations usually require, in addition, some mathematically defined functions of one or more variables. In fact, even the linear algebra of Section **c** usually leads to some square-root operations, requiring an algorithm for this function.

The problem then is to develop computer algorithms to approximate a mathematical function for a specified range of its argument(s). For most applications the approximating algorithm is assumed to use only the basic arithmetic operations, without calls to other algorithms. Therefore, the approximating functions will usually consist of one or more polynomial or rational functions. The mathematical and computational work on such approximations is extensive, particularly for functions of a single variable. This and the two following sections discuss some of the currently more important methods, emphasizing operational requirements and basic mathematical results.

By far, the most work and the most satisfactory results apply to the case of a real-valued function of a single real variable, say f(x), over a given interval of values for x. Approximation of f(x) is usually viewed as a two-stage process. First, one defines an approximating function, h(x). Second, a computer algorithm to evaluate h(x), say H(x), is written. The function, f(x), is presumably expensive to evaluate, whereas h(x) is less so, usually involving only basic arithmetic operations.

The central requirements for approximation are accuracy, speed and size. The goal is to find the fastest, smallest program returning values, H(x), differing as little as possible from f(x). Fixing a criterion for one of the goals generally causes the other two to work against each other: for a given accuracy, one can usually obtain a faster algorithm by making the program larger, and so on. The choices made in practice are always somewhat arbitrary, particularly when the algorithm is to be applied to a wide variety of problems.

Two important forms of approximation are: approximation by a single rational function, particularly using maximum error as a criterion, and by piecewise polynomials, particularly the class of smooth functions usually called *splines.* The first class underlies many standard mathematical routines, and is particularly appropriate for library functions of a single variable. For less standard cases or for other criteria of approximation, the second class shows considerable promise.

The accuracy of an approximation, H(x), can be measured in several ways. Most analyses are in terms of the *absolute error function,*

$$e_a(x) = f(x) - H(x) \tag{33}$$

or the *relative error function*

$$e_r(x) = (f(x) - H(x))/f(x). \tag{34}$$

Which of these, if either, best describes a particular situation depends upon the behavior of f(x) and the use of the results. While e_r is independent of rescaling for f(x), assuming H(x) is also rescaled, it is undefined or infinite when f(x) = 0. Mathematically, we can define a general weighted error function,

$$e(x) = w(x)(f(x) - H(x)) \tag{35}$$

and use it in discussions to stand for e_a, e_r, or several other choices of error function.

Notice that e(x) may be viewed as composed of two parts:

$$
\begin{aligned}
e(x) &= w(x)\Big[(f(x) - h(x)) + (h(x) - H(x))\Big] \\
&= t(x) + r(x). \qquad (36)
\end{aligned}
$$

The two parts are usually called the *truncation error,*

$$t(x) = w(x)(f(x) - h(x)), \qquad (37)$$

which is the difference between f(x) and the mathematical approximation derived for it, and the *rounding error,*

$$r(x) = w(x)(h(x) - H(x)), \qquad (38)$$

which is the error incurred in evaluating the algorithm, given the approximation. Neither name is entirely appropriate (approximation error and evaluation error would probably be better), but the terminology is standard by now. There is, however, an essential ambiguity, often overlooked. If we define h(x) by a mathematical criterion, some calculations are required to determine the resulting function, and these calculations always involve error. Thus t(x) could be further decomposed into error inherent in the definition of h(x) and error induced by having to determine h(x) numerically (for example, in determining the coefficients of a rational approximation).

Most of our discussion is of truncation error, including both of its components. Assuming we do not know a fixed set of x for which H(x) is required, some measure of error over the entire range is required, similar to the measures of the size of error matrices in Section **c**. A mathematically attractive measure is the maximum error:

$$\max|e(x)|,$$

for all x in the region. This leads to a widely used form of approximation: namely, minimax approximation by a single rational or polynomial function.

Specifically, suppose we write a general rational function

$$h(x) = p(x)/q(x) \qquad (39)$$

where p(x) is a polynomial of degree m and q(x) is a polynomial of degree n, say. We will always assume that h(x) is *irreducible,* meaning that no polynomial is an exact factor of both p(x) and q(x). The central mathematical result of minimax approximation is:

Theorem: Suppose t(x) is the truncation error defined in (37) and $H^{m,n}$ is the set of all irreducible rational functions h(x) as in (39). Then

(1) There exists a unique rational function $h^*(x)$ of degree (m',n') where $m' \leqslant m$, $n' \leqslant n$ such that h^* minimizes the maximum truncation error; i.e., if $t^*(x)$ is t(x) when $h = h^*$,

$$\mu^* = \max_{[a,b]} |t^*(x)| = \min_{h \epsilon H^{m,n}} \max_{[a,b]} |t(x)|; \tag{40}$$

and

(2) The error function, $t^*(x)$, takes its extreme values at points

$$x_1^* < x_2^* \cdots < x_N^*$$

with equal magnitude and alternating sign; i.e.,

$$t^*(x_i) = (-1)^{i-1} t^*(x_1)$$

$$|t^*(x_i)| = \mu^* \tag{41}$$

The property (41) is called *equal-ripple* behavior, and is the critical property used to compute minimax approximations.

There exists a great deal of mathematical discussion of this result, its special cases and its extensions. The reader wishing to see a fairly thorough, but readable, treatment of the Theorem may consult Ralston (1967, Chapter 5). For an exhaustive discussion of many mathematical questions about approximation, the two volumes by Rice (1964;1969) are recommended. For our purposes, the more delicate theoretical issues can be glossed over. We assume the simplest, and by far the most common case; namely, that the approximation is *nondefective* and the error curve is *standard.* By this is meant that the degrees of the numerator and denominator are exactly m and n, and the number of relative extrema of $t^*(x)$ is exactly $m + n + 2$.

Different methods for determining minimax approximations vary according to the form of h(x) (general rational or polynomial), to whether the approximation is over an interval or a discrete set of points and to which algorithm is used to generate an approximation. We begin by describing the use of *Remes'* algorithm for a general rational approximation over an interval and then discuss other possibilities briefly.

Remes' algorithm is based on the following observations. First, if $h_k(x)$ is any rational function of degree (m,n), let

$$x^*_{k,1} < x^*_{k,2} < \cdots < x^*_{k,N}$$

be the relative extrema of the error function $t_k(x)$ in (37) for approximating $f(x)$ by $h_k(x)$. Then, for any $h_k(x)$,

$$\max_i |t_k(x^*_{k,i})| \geqslant \mu^* \tag{42}$$

where μ^* is the minimax error in Theorem 1. If equality holds, we have found the minimax approximation. Secondly, given now the points $x^*_{k,i}$ with $N = m + n + 2$, we can usually solve the equations

$$w_k(x^*_{k,i})\left[f(x^*_{k,i}) - h_{k+1}(x^*_{k,i})\right] + (-1)^i \mu_{k+1} = 0 \tag{43}$$

for the $m + n + 1$ coefficients needed to determine the rational function $h_{k+1}(x)$ and the error μ_{k+1}. The iterated use of these two steps produces the algorithm.

Remes' Algorithm

I. Determine an initial estimate for the extrema, x^*_{0i}. Set $k = 0$.

II. Repeat through Step V.

III. If the approximation is close enough to the minimax, exit.

IV. Solve (43), approximately, for the next approximation, and for μ_{k+1}.

V. Determine, approximately, the extrema, $x^*_{k+1,i}$, of the new approximation.

There exist a number of published algorithm for this procedure, notably Cody, Fraser and Hart (1968) or Werner, Stoer and Bommas (1967) for rational approximation on an interval.

Many variations in the implementation of this algorithm are possible, and the detailed coding has considerable effect on both the speed of convergence in well-behaved problems and the ability to solve very difficult problems. Variations are possible in the methods used to approximate the solutions to the nonlinear problems of

Steps IV and V, and on how close an approximation is desired. The details are beyond our scope: the reader is referred to the algorithms above and to the discussion in Rice (1969, Section 9-7).

The algorithm as stated applies not only to rational approximation on an interval but to the other minimax problems as well, with the individual steps suitably redefined. If $f(x)$ is defined on an explicitly discrete set, then the extrema of the error curves, $t_k(x)$, will be taken over the discrete set.

The specialization of the approximating function to a polynomial gives a somewhat simpler approximation problem. The approximating function now depends *linearly* on the coefficients, so that the determination of $h_k(x)$ is a simpler problem. A thorough algorithm for minimax polynomial approximation on an interval is given by Golub and Smith (1971). There are, however, two arguments suggesting that this form of approximation is not very attractive. First, there is considerable empirical experience to suggest that rationals with $n > 0$ (often with numerator and denominator of about the same degree) give better minimax approximations than a polynomial with the same number of terms (and therefore costing about the same to evaluate). The book by Hart et al. (1968) tabulates many examples. Second, Powell (1967) showed mathematically that polynomial approximations on an interval obtained by interpolation on a fixed set of values or by minimizing the integrated squared error can *always* be made nearly as good, in the sense of maximum error, as true minimax approximations.

Therefore, one is led to conclude that, if true minimax approximation of a smooth function is desired, one should use rationals. For polynomial approximations, simpler approximating methods are generally adequate. A more attractive alternative to rational functions for most applications are the piecewise polynomials or *splines,* discussed in Section **e**.

Remes' algorithm has been the most widely used procedure for minimax rational approximation. There has been some interest in other procedures, however. One such is called the *differential correction* algorithm. Barrodale, Powell and Roberts (1972) present a version of this algorithm for approximation on a discrete set, say on the points $x_1, x_2, \ldots, x_N$. Suppose at the kth iteration, the approximating function is

$$h_k(x) = p_k(x)/q_k(x)$$

and the maximum error is μ_k. The differential correction algorithm then chooses h_{k+1} to minimize

$$\max_{x_i} \left(|f(x_i)q_{k+1}(x_i) - p_{k+1}(x)| - \mu_k Q_{k+1}(x_i) \right) / w_k(x_i) \qquad (44)$$

subject to

$$Q_{k+1}(x_i) > 0$$

for all i. The weighting function $w_k(x)$ is chosen as $Q_k(x)$ by Barrodale et al. (1972). This algorithm seems an attractive alternative to Remes' algorithm on theoretical grounds. It is guaranteed to converge, its final rate of convergence is quadratic, and (for discrete sets) it is easy to program. Further practical experience and readily available algorithms may increase its popularity.

e. Spline Approximations

The approximating functions receiving the most attention as alternatives to rational functions are smooth piecewise polynomials. Specifically, the approximating functions, h(x), are constrained to be polynomials of chosen degree, d, on the set of intervals $[z_i, z_{i+1}]$, where

$$a = z_0 \leqslant z_1 \quad \cdots \quad \leqslant z_{N+1} = b \qquad (45)$$

and h(x) and some chosen number of its derivatives are required to be continuous through the interval [a, b]. The points z_i are frequently called the *knots.*

The special case of greatest interest has been that of *spline* functions, in which (d-1) derivatives are constrained to be smooth. These functions have generated much mathematical and practical interest and have been applied to a variety of other problems besides approximation; for example, integral and differential equations, integration, and the calculus of variations. For a general introduction, see the book by Ahlberg, Nilson and Walsh (1967), and Chapter 10 of Rice (1969). The terms *spline* and *knot* refer to jointed rubber, wooden, or metal strips used in nineteenth century engineering to follow the shape of an object for drawing or calculation. Spline functions are a numerical analog of this process.

Several methods for representing splines are available. The relevant criteria, when *evaluating* the spline are accuracy and speed. On the other hand, for *estimating* the coefficients defining a particular approximation, the effect of the representation on the numerical conditioning of the estimation will be important. Nearly all practical representations amount to a two-stage process for evaluation: first, determine the subinterval containing the given x; second, evaluate h(x) by a special formula for this subinterval. Another way to say this is that a *local basis* for the spline functions is defined; that is, a set of functions, say $P_j(x)$, having the property that

$$P_j(x) \neq 0 \quad \text{only if} \quad z_L < x \leqslant z_R \tag{46}$$

with the knots z_L and z_R depending only on which subinterval contains x, for given j. See Section **f** for details. Then if we represent h(x) as

$$h(x) = \sum_j v_j P_j(x)$$

it follows that the evaluation of h(x) is in fact of the form

$$h(x) = \sum_{P_j(x) \neq 0} v_j P_j(x) \tag{47}$$

which in general simplifies both evaluation and estimation.

The simplest such representation is to choose a basis for the polynomials of degree d on each subinterval, $[z_i, z_{i+1}]$, and then take P_j to be one of these polynomials on the subinterval and zero outside. This makes h(x) easy to evaluate but turns out to give difficult estimation problems. Other definitions of $P_j(x)$ have been given; one attractive compromise is the so-called *B-splines* due to Schoenberg. For a discussion of their properties and practical use, see Section **f**, and de Boor (1972). For algorithms, see the Appendix.

From the point of view of approximation, the spline functions have three distinct advantages over rationals. First, their piecewise definition leaves them freer geometrically to follow the shape of a curve. Second, splines can be applied with a variety of fitting criteria and side conditions to obtain approximating curves with specific properties. Third, splines have various optimality properties of interest (some of which are tangential to the approximating problem itself). We do not discuss these in detail. Rice (1969, 154-159) gives a centrally important mathematical theorem, which is a generalization of a number of results. A more specialized and less precise statement is as follows.

Let L be a linear functional: i.e., an operator acting on a *function* to return a real number, and satisfying linearity,

$$L(ah_1 + bh_2) = aL(h_1) + bL(h_2)$$

for scalars a, b and functions h_1, h_2. Examples are integrals and derivatives at a point. The class of functionals involved is defined by

$$L(f) = \sum_{r=0}^{k-1} \int_a^b (d^r f/dx^r)\, d\mu_r(x) \tag{48}$$

where μ_r is allowed to be any function of *bounded variation.* The essential result is that the h for which L(h) is the best approximation to L(f) among a certain class of approximations (roughly, linear combinations of functions satisfying $L_i(h) = \alpha_i$ for chosen functionals, L_i, and values, α_i) is given for h a suitably chosen so-called *natural spline.* Natural splines are defined by taking $a = -\infty$, $b = \infty$, requiring h(x) to be of odd degree d = 2k-1 in all intervals and to be of degree k in the intervals $[-\infty, z_1]$ and $[z_N, \infty]$. Some special cases of this result are the interpolating and smoothing splines.

For whatever approximating problem splines are desired, the dominant fact in mathematical and computational consideration is that h(x) is a *linear* function of the coefficients in (47), given the knots, and a *nonlinear* function of the knots themselves. For fixed knots, the determination of the desired spline is usually easy; for arbitrary knots, it tends to be difficult. The following are a few of the many applications of splines, along with references to algorithms, when available. See the Appendix for further algorithms.

(1) *Interpolation.* There are a number of formulations of the interpolation problem. Given an arbitrary set of n points, (x_i, y_i), some specialization of the definition of the spline function is needed to produce a specific result. A widely used form is to require h(x) to be a natural spline and to take N = n internal knots at $z_i = x_i$. This results in a simple set of linear *band equations;* i.e., linear equations with, in this case, only the diagonal elements and k elements to the left and right of the diagonal nonzero. A system of p such equations can be solved in $O(kp^2)$ operations as compared to $O(p^3)$ more generally (see Section **5.g**). Herriot and Rheinsch (1973) give ALGOL60 algorithms.

(2) *Smoothing.* There are two distinct cases here: smoothing with or without a requirement of interpolation as well. The two problems are closely related to each other, at least if the norms used to measure deviation and smoothness are based on squared values. Frequently, the smoothness of H(x) is measured by

$$J = \int_a^b (d^k h/dx^k)^2\, dx; \tag{49}$$

i.e., by the integral of the squared k^{th} derivative of $h(x)$, assuming again that $h(x)$ is a spline of odd degree $2k-1$. The total squared deviation of the spline from some chosen n points (x_i,y_i), weighting the ith point by $w_i \geqslant 0$ is

$$E = \sum_{1}^{n} w_i(y_i - h(x_i))^2 \tag{50}$$

Smoothing with interpolation can be defined as

$$\text{Minimize } J \text{ subject to } E = 0. \tag{51}$$

This problem, given that the knots are chosen at the interpolating points and $n = N$, as above, is in fact solved by the natural spline interpolating function. This is a special case of the optimality result on linear functionals mentioned above. The solution to the problem

$$\text{Minimize } J + \alpha E \tag{52}$$

for fixed chosen α also leads to a set of linear band equations and is also a special case of the optimality results on natural splines. Woodford (1970) gives an ALGOL60 algorithm for both these problems.

(3) *Least Squares.* Now suppose we drop the explicit smoothing requirement, and simply look for spline regressions. *If* the knots are fixed, this leads to a linear least-squares problem, possibly with linear equality constraints. If the representation (47) in terms of a local basis is used, then the matrix A in the least squares problem such that

$$A[i, j] = P_j(x_i)$$

will be sparse; i.e., many of the elements of A will be known a priori to be zero.

If the elements of x are ordered, there is generally a natural ordering of the basis splines as well, resulting in a matrix A with a special kind of sparseness:

$$A = \begin{bmatrix} A_1 & & 0 \\ 0 & A_2 & 0 \\ 0 & & A_{N+1} \end{bmatrix} \tag{53}$$

where A_j is an n_j by p_j matrix, n_j being the number of data points in (z_j,z_{j+1}). Notice that, in general, the left columns of A_2 will overlap A_1 and so on. The exact pattern will depend on the choice of basis, and there may be further sparseness within A_j. One will want to take advantage of this structure when doing least squares or other linear operations on A. For example, the Givens decomposition (see Section **5.c**) allows us to skip over zero elements when forming an orthogonal decomposition of A. An alternative procedure is to form the sum of squares matrix, which will be of band form, and carry out a Choleski decomposition (see Section **5.c** and Cox (1971)). For ill-conditioned problems, iterative improvement (Section **5.f**) can be applied, again taking advantage of (53).

For the case of *free* knots, there may not be a unique solution. However, local solutions can be searched for, treating the linear solution for v, given the knots, as an algorithm. Suppose the solution for fixed knots is written as $h^*(x{:}z)$ with z the vector of knots. Then to find z^* to minimize

$$\sum w_i(y_i - h^*(x{:}z))^2$$

is a nonlinear least squares problem. Those methods discussed in Section **6.g** that do not require derivatives would be attractive. However, for a sizeable number of knots, the high cost and possible nonuniqueness of the solution make this an unattractive problem.

(4) *Other Methods.* Methods (1) to (3) are the most important practical spline approximations. Splines can be found to fit functions by minimizing the sum of absolute deviations:

$$\sum |y_i - h(x_i)|.$$

For fixed knots, this becomes a linear programming problem (see Section **6.i**) using the representation (47). The feeling that this criterion may be more *robust,* i.e., less affected by gross errors or a non-normal distribution for the data, makes the solution attractive, but little practical experience has been reported. Again, computational complexity and nonuniqueness plague the free-knot case.

Minimax spline approximation has received some mathematical attention, but is not very attractive in practice. Even in the fixed-knot case, nonuniqueness is possible (Schumaker, 1968; Rice, 1969, Section 10.5). In the free-knot case, even the characterization of the best approximations, by a form of the equal-ripple argument, does not entirely characterize solutions (Braess, 1971).

Splines can also be attractive for certain forms of constrained optimization. For example, a requirement that h(x) be *monotone* at the knots, i.e., that the derivative there be nonnegative, introduces linear inequality constraints on the coefficients. Such applications of splines have considerable interest, but have not yet been much explored. See the comments on B-splines, in Section **f**, (61) to (65).

f. Evaluation of Approximations

Let us now turn to the *evaluation* of an approximation, $h(x)$, by means of an algorithm, $H(x)$. Two sources of error are relevant in this evaluation, analogous to the rounding error and numerical instability discussed in Section **c** for the basic arithmetic operations. First, on the assumption that x is exact, the *rounding error* (38) in evaluating $h(x)$ can be estimated for any chosen scheme of evaluating an approximation. Second, on the more realistic assumption that x differs by ϵ, say, from its true value, the resulting propogated error in $H(x)$ can also be studied.

The second of these depends primarily on the function being approximated, assuming that the first source of error has been reasonably controlled. Rounding error depends upon the form in which $h(x)$ is represented and also, in some cases, upon the specific coefficients. The choice of representation and evaluation method must usually balance good rounding error properties against an increased number of arithmetic operations: for most library routines, it does not seem justifiable to risk much rounding error for the sake of, generally, minor savings in computing time.

It is convenient to write most evaluation algorithms as iterations returning successive partial results, $H_0, H_1, \cdots, H_N$. Each H_i is the result of a simple calculation depending on x, a few coefficients, and a few previous H_j terms. Consider an ordinary polynomial,

$$p(x) = \sum_{i=0}^{m} a_i x^{m-i} \tag{54}$$

The following iteration, known as *Horner's method,* can be seen to produce the value of $p(x)$ as H_m:

$$\begin{aligned} H_0 &= a_0 \\ H_{i+1} &= a_{i+1} + x H_i \end{aligned} \tag{55}$$

Notice that it is not necessary to store the various partial results in this scheme, as they are each used only once. In fact, more efficient code may be generated, for known values of a_j, by writing the evaluation in a single nested form:

$$H_m = (a_m + x(a_{m-1} + x(\cdots (a_1 + x a_0) \cdots)$$

This produces the same arithmetic operations, but not in the same order. Horner's method is very commonly used and represents a compromise between the fastest and the most accurate methods.

The powers, x^i, have poor conditioning on most intervals. Basically, this says that x^i looks very much like a linear combination of lower powers of x; more precisely, a matrix of values of $1, x, x^2, \cdots, x^m$ at points spread over an interval will tend to have a large *condition number* (see Section **5.f**). To some extent, this increases the rounding error in (55). To improve accuracy, one may write p(x) in terms of a set of orthogonal polynomials: i.e., a set of $m+1$ polynomials, $q_0(x), \cdots, q_m(x)$, with $q_j(x)$ of degree j and with an orthogonality property:

$$\sum_{k=0}^{M} w_k q_i(z_k) q_j(z_k) = 0, \quad i \neq j \tag{56}$$

for given weights, w_k, and points, z_k. The most important set for approximation are the *Chebyshev polynomials,* defined on the interval $[-1, 1]$ by

$$q_0(x) = 1, \; q_1(x) = x$$

$$q_{i+1}(x) = 2x\, q_i(x) - q_{i-1}(x) \tag{57}$$

These useful and interesting polynomials are related to trigonometric functions and to minimax approximation. For a full discussion, see the book by Clenshaw (1962). The definition of $q_i(x)$ leads to a formula similar to Horner's method. If p(x) is written as a linear combination of the q_i:

$$p(x) = \sum_{i=0}^{m} b_i\, q_i(x);$$

then an iteration can be defined by

$$H_0 = 0; \quad H_1 = 0;$$

$$H_{i+1} = 2xH_i - H_{i-1} + b_{i-1} \tag{58}$$

In this case p(x) is not precisely the last H_i; instead,

$$p(x) = ½(H_{m+1} - H_{m-1})$$

For details, see Clenshaw (1962). Use of this form may be indicated if the error bounds on (55) are unattractive. These bounds may be computed by an iteration similar to (55) itself:

$$\delta_{i+1} \leqslant |x|\,\delta_i + (|x\,H_i| + |H_{i+1}| + |a_{i+1}|)\,\epsilon \tag{59}$$

where ϵ is the relative precision; specifically

$$\epsilon = 2^{-t},$$

with the assumptions of Section **b**,. See Hart et al. (1968, Section **4.5**). If the Chebyshev form is to be used, one must transform the a_i coefficients to the equivalent b_i form. A procedure for this is outlined in Hart et al. (1968, Section **4.7**). Other representations are possible, giving either speed or accuracy, but these two seem good choices: Horner's method as long as it is not too inaccurate and the Chebyshev form otherwise.

Rational functions may be evaluated as the ratio of two polynomials. For most purposes this is an acceptable solution; it is very simple and will be accurate if the polynomial evaluations are accurate. Some special techniques for evaluating rationals are described in Hart et al. (1968, section **4.6**). These may sometimes be more accurate than Horner's method and cheaper than the Chebyshev form. See the reference for details.

Bounds on rounding error assume that x is given exactly, but in practice it will be the result of previous computation and/or dependent on observed values. In either case, one can assume that, instead of computing H(x) we compute H(x + e) where e is the error in x. An understanding of the resulting error in the approximation to f(x) may be critical, if the result of a computation is to be accurate. The problem can be formulated in terms of relative or asbolute error in x, the former being natural for computational error. For example, suppose we wish to compute exp(x) but are given $x \times (1 + e)$. Then, ignoring errors or approximation, we get as a result $\exp(x)(1+\delta)$ where the error, δ, is given by

$$\delta = \exp(x\,e) - 1$$

Clearly, as x becomes large, so does δ.

This sort of error does not come from approximating difficulties but reflects an inherent sensitivity in the function. Notice that the function need not become large. Similar problems exist in trigonometric functions:

$$\sin(x(1 + e)) = \sin(x)\cos(x\,e) + \cos(x)\sin(x\,e)$$

When x becomes large enough that xe is not nearly zero the propogated error will be significant.

A related problem may occur if the propogated change in f(x) due to a change in x is very *small.* Consider, for example, the expression

$$\cos(Ax) - B\cos(Cx) \tag{60}$$

where A, B, and C are known coefficients and x ranges on an interval including zero. While this expression looks eminently reasonable, the relative error in the result will be arbitrarily large as x tends to zero, regardless of the accuracy with which we evaluate the cos(x) function. For, near zero, the Taylor series for cos(x) gives

$$\cos(x) = 1 - \frac{x^2}{2} + O(x^4).$$

If x is small then nearly all the accuracy in x^2 is lost when it is subtracted from 1; indeed, if $|x| < 2^{-1/2\,t}$, $\cos(x) = 1$. But this does *not* mean that we cannot compute (60) accurately. If we use the trigonometric identity

$$\cos(x) = 1 - 2\sin(x/2)^2$$

then this result becomes

$$2B\sin(Bx/2)^2 - 2\sin(Ax/2)^2 + (B-1)$$

The first two terms are both small and can be computed to high relative accuracy. The last term is independent of x; its accuracy depends on whether we can find B to high precision, but in any event the result is much less catastrophic than (60). This argument can be applied generally in evaluating trigonometric expressions. All trigonometric functions of x can be evaluated in terms of the single function, tan(x/2), so that rearrangement can be done without serious increase in cost (see Hart et al. (1968, p. 114)).

For the evaluation of *spline approximations,* the choice of representation in (47) is of primary concern. Generally one is likely to have estimated (47) using a basis such as the B-splines, incorporating the continuity conditions on splines and producing straightforward, well-conditioned estimation problems. One may then evaluate the approximation using the same basis, or transform the coefficients to those for a set of ordinary polynomials on the subintervals. The latter will generally provide faster evaluation, at the cost of increasing the number of coefficients.

Let us consider specifically the B-splines. Although it is not necessary, the assumption that the knots z_i in (45) are *distinct* will simplify discussion. Let $N_{i,k}(t)$ stand for the *normalized B-spline* of degree k corresponding to the ith subinterval, for $i = 0, \cdots, N-b+2$.

It is convenient to define $N_{i,k}(t)$ recursively as follows. Initially,

$$N_{i,1}(x) = \begin{cases} 1 & \text{if } z_i \leqslant x < z_{i+1} \\ 0 & \text{else} \end{cases} \tag{61}$$

and then, given degree $0, \cdots, k-1$,

$$\begin{aligned} N_{i,k}(x) &= (x - z_i)N_{i,k-1}(x)/(z_{i+k-1} - z_i) \\ &\quad + (z_{i+k} - x)N_{i+1,k-1}(x)/(z_{i+k} - z_{i+1}) \end{aligned} \tag{62}$$

This is usually derived as a property of B-splines, but here it serves as a definition. Then $N_{i,k}(x)$ is a *strictly positive* polynomial of degree $k-1$ on $[z_i, z_{i+k}]$ and 0 elsewhere, as can easily be proved by recursion. The positiveness is a very useful property in several ways. For example, it allows us to generate positive, monotone, or convex approximating functions simply by constraining the coefficients in the approximation of $f(x)$ or its first or second derivative to be positive.

Now suppose we have produced an approximation in terms of the B-splines of degree d. If $z_i \leqslant x \leqslant z_{i+1}$, the application of (47) gives

$$H(x) = \sum_{j=i-d+1}^{i} a_j N_{j,d}(x). \tag{63}$$

There are several mechanisms for evaluating this approximation (de Boor, 1972). One simple and flexible scheme is just to form the set

$$((N_{j,k}(x), j = i{-}l{+}1, \cdots, i), k = 1, \cdots, d)$$

directly from the recursive definition (62). The dth outer loop of the list produces the numbers

$$(N_{i-d+1,d}(x), \cdots, N_{i,d}(x))$$

which generate $H(x)$ from (63). Normally $N_{j,k}(x)$ can overwrite $N_{j-1,k-1}(x)$.

An additional advantage of the representation in (63) is that derivatives of $H(x)$ can be computed as well. For,

$$dH(x)/dx = \sum_{j=i-d+2}^{i} a_j^{(1)} N_{j,d-1}(x) \tag{64}$$

where the coefficients $a_j^{(1)}$ are given by

$$a_j^{(1)} = (d-1)(a_j - a_{j-1})/(z_{j+k-1} - z_j) \tag{65}$$

This can be verified by differentiating the basic recursion and substituting in (63). As noted in Section **e**, spline approximations are well-suited to estimation of derivatives. An estimate of $df(x)/dx$ follows from (64), by using the second last set of $N_{j,k}(x)$. Thus derivative estimation is nearly free. See the Appendix for algorithms using B-splines.

g. Numerical Integration

Suppose $f(x)$ represents a real function of a single real variable. The problem of *numerical integration* is then to compute

$$J(a, b, f) = \int_a^b f(x)\, dx \tag{66}$$

There are three major classes of algorithms to approximate J: *approximation techniques* using a fixed number of values, *automatic methods* using approximations iteratively, and *Monte-Carlo methods.* The third class is applied mostly in the case of multivariate integration and is discussed in Section **7.i**.

The oldest approximation techniques are based on interpolating a sequence of points, (x_i, y_i), where $y_i = f(x_i)$,

$$a = x_0 < x_1 \cdots < x_N < x_{N+1} = b$$

and where the x_i are equally spaced. The best-known examples are: interpolation by straight lines (the trapezoidal rule),

$$\sum_{i=0}^{N} \tfrac{1}{2}\, (y_i + y_{i+1})\, (x_{i+1} - x_i),$$

which reduces when $x_{i+1} - x_i = \delta$, for all i, to

$$J_T = \delta(\tfrac{1}{2}\, (y_0 + y_{N+1}) + \sum_{i=1}^{N} y_i) \tag{67}$$

and interpolation by quadratics (Simpson's rule), for the case that N is odd, say $N = 2K + 1$,

$$J_S = (\delta/3)(y_1 + 4 \sum_{j=1}^{K} y_{2j} + 2 \sum_{j=1}^{K-1} y_{2j+1} + y_N) \tag{68}$$

again assuming that $x_{i+1} - x_i = \delta$, for all i.

Equations (67) and (68) are based on evaluating $y_i = f(x_i)$ at the end-points of *equally spaced* subintervals and then approximating $f(x)$ by a *low-order* polynomial approximation on each subinterval.

Other integration formulas arise from allowing the evaluation at specially chosen unequally spaced points and/or approximating $f(x)$ by a high-order approximation. Assume for the moment that we have made the linear transformation to $(x - (a + b)/2)/(b - a)$ which reduces (66) to the interval $[-1, 1]$.

The use of unequally spaced points leads to *Gaussian quadrature* techniques. Here we allow $x_1,\ldots,x_N$ to be free parameters and attempt to evaluate the integral by

$$J_G = \sum_{i=1}^{N} c_i f(x_i) \tag{69}$$

Having the x_i at our disposal, we can hope to make J_G more accurate, using a given number of values of $f(x)$. Specifically, one can show that (69) will be exact, ignoring rounding error, for any polynomial of degree $2N-1$ or less if:

1. The points x_i are the zeroes of the polynomial $p_N(x)$ defined by the orthogonality requirement

$$\int_{-1}^{1} p_i(x) p_j(x)\, dx = \begin{cases} 1 & i = j \\ 0 & i \neq j \end{cases} \tag{70}$$

2. The coefficients c_i are given as the integral of the polynomial defined as 1 at x_i and 0 at x_j for $j \neq i$.

The polynomial defining c_i is the elementary Laplace polynomial

$$L_i(x{:}x_1 \ldots x_N) = \prod_{j \neq i} (x - x_j) / \prod_{j \neq i} (x_i - x_j). \tag{71}$$

For a derivation, see Hildebrand (1956, 319-321). The Gaussian results are often used in a weighted form. To evaluate (66) we approximate instead

$$J(a, b, f_*, w) = \int_a^b f_*(x) w(x)\, dx \tag{72}$$

for $f_*(x) = f(x)/w(x)$ and chosen weighting function $w(x)$. The only change in the calculations is that all integrals now have $w(x)\,dx$ instead of dx. Common weighting formulas include trigonometric or exponential functions, sometimes with a change of variable to facilitate integration on other intervals, such as $[0,\infty]$. See Hildebrand (1956, 323-351).

Higher-degree polynomial approximation with *fixed* step size produces the general class of *Newton-Coates* quadrature formulas

(Hildebrand, 1956, 71-84). For modern computation, these formulas are mainly used in the context of automatic algorithms, to be discussed below.

Another class of approximating functions relevant to integration is based on the *Chebyshev polynomials* (see (57) in Section **f**). These are derived by making the change of variable

$$x = \cos\theta \tag{73}$$

and interpolating f(x) on the roots of the polynomial

$$C_N(x) = \cos(N\theta) \tag{74}$$

i.e., on the points, θ^{*j}, such that $N\theta^{*j} = \pi j$, for j=0,...,N. These polynomials are related to minimax approximation in that $T_N(x) - x^N$ is the minimax approximation to x^N of degree N.

Powell (1967) showed that interpolating on θ^{*j} is nearly as good an approximation as a minimax polynomial (see Section **d**). Another application of the Chebyshev polynomials is to approximate f(x) by the finite cosine series,

$$f_c(\theta) = \sum_{j=0}^{N}{}'' a_j \cos(j\theta). \tag{75}$$

The notation $\sum''$ means that first and last terms are multiplied by ½. The coefficients a_j are defined by the finite cosine transform

$$a_j = 2\sum_{i=0}^{N}{}'' f(\cos(\theta^{*i}))\cos(j\theta^{*i}), \tag{76}$$

with the above definition of θ^{*i}. The computation of a_j is done by the *Fourier transform* technique below.

The application of (75) to integration comes from changing variables (again assuming [a, b]=[−1, 1]):

$$J(a, b, f) = \int_0^{\pi} f(\cos\theta)\sin\theta d\theta \tag{77}$$

If $f(\cos\theta)$ is approximated by $f_c(\theta)$ then the integration of f_c gives

$$J_c = \int_0^{\pi} f_c(\theta)\sin\theta\, d\theta = \left(\frac{2}{N}\right)\sum_{j=0}{}'' a_j/(1-j^2). \tag{78}$$

The integration techniques have all been described so far as if a specific formula and fixed set of function values were prechosen.

In practice the goal is nearly always to approximate the integral to a specified accuracy. Clearly, this is not possible by a fixed formula. Most modern integration algorithms operate iteratively, using a succession of formulas of increasing refinement chosen from one or more of the families above, until the desired accuracy is judged to have been reached or until the algorithm has failed, say, by computing more than the allowed number of function values.

Such algorithms are generally called *automatic* quadrature methods. Nearly all the methods above generate some form of automatic procedure. These divide further into *whole-interval* or *adaptive* methods. Whole-interval methods follow the simple strategy just outlined, refining the approximation of J(a, b, f) until completion. Adaptive methods permit in addition the subdivision of the integral; e.g.,

$$J(a, b, f) = J(a, \tfrac{1}{2}(a+b), f) + J(\tfrac{1}{2}(a+b), b, f).$$

If the current approximations on the integral on *both* subintervals are adequate the algorithm is finished. Otherwise the refinement process is applied, recursively, to one or both of the subintervals.

Any automatic integration routine has two main problems:

(i) computing successive refinements efficiently;

(ii) estimating accuracy.

For (i), the algorithm should make maximum use of computed function values; specifically, the refinement should reuse the values of f(x) in the current formula. This can be done without wasting any function values by formulas like J_T and J_S which use equally spaced points. Other methods will waste at least some values on subdividing. Whole-interval automatic procedures are available for all methods.

The Appendix lists some of the currently available automatic integration algorithms. Details on the methods are available in the references. Aside from differences due to the quality of coding, there is an inherent balance in automatic integration between high efficiency on easy problems and good behavior on hard problems. For one thing, the choice of a test for convergence is likely to reflect the degree of conservatism applied.

The adaptive algorithm DQUAD (successful on many difficult problems) and the nonadaptive Clenshaw-Curtis have been used successfully on a wide range of problems at Bell Laboratories.

h. Fourier Transforms; Spectral Analysis

The *spectral analysis* of time-series data is a widely useful technique. It arises, for example, if one wishes to model data as a set of periodic components or, less formally, if one wishes to explore possible periodic effects. Spectral analysis and associated statistical techniques are closely related to the numerical topic of *Fourier transformation.* For an introduction to the statistical background, see Jenkins and Watts (1968) or Bloomfield (1976). For a more detailed discussion of the Fourier transform, including some computational detail, but not from the viewpoint of data analysis, see Brigham (1974).

For purposes of computational discussion, it will be convenient to assume the existence of *complex numbers,* as pairs of reals,

$$\textbf{complex}\ z=(\ \textbf{real}\ a,\ \textbf{real}\ b) \tag{79}$$

with the usual rules of complex arithmetic; in particular,

$$(a_1, b_1) \times (a_2, b_2) = (a_1a_2 - b_1b_2,\ a_1b_2 + a_2b_1).$$

We reserve the letter i for the complex number

$$i=(0,1).$$

If z(t) represents a complex function of the real variable, t, the *continuous Fourier transform,* $\phi(u)$, is a complex function of a real variable, u, defined by

$$\phi(u)=\int_{-\infty}^{\infty} z(t)\ \exp(i2\pi ut)\ dt. \tag{80}$$

The continuous transform has also the important inverse formula

$$z(t)=\int_{-\infty}^{\infty} \phi(u)\ \exp(-i2\pi tu)du. \tag{81}$$

An important class of applications of Fourier transforms is to the relation between probability densities and characteristic functions. Here z(t) is a (real) density and $\phi(u/2\pi)$ is the characteristic function. By replacing z(t) dt by the probability element dF(t) a much more general mathematical basis is developed. Readers are referred to Feller's (1971, Chapter XV) elegant, idiosyncratic treatment.

For computational purposes, we must work with discrete quantities, even if the continuous transform is our conceptual model. The basic tool is the *finite* or *discrete* Fourier transform of a complex

N-vector, z; namely, the complex N-vector, ϕ:

$$\phi[u]=\sum_{t=0}^{N-1} z[t] \exp(i2\pi tu/N) \tag{82}$$

(Notice that in this section we follow the convention of indexing vectors on $0, \cdots, N-1$; translation to the range $1, \cdots, N$ is obvious.) The discrete transform also has the inverse formula

$$z[t]=N^{-1}\sum_{u=0}^{N-1} \phi[u] \exp(-i2\pi ut/N). \tag{83}$$

The near-symmetry of (82) and (83) leads to a number of possible conventions about which is the direct and which the inverse transform and how to divide the scaling factor, N^{-1}, between the two forms. The practical effects are minor, so long as one knows which convention is being followed. Readers needing clarification on the meaning of terms are referred to Rabiner et al. (1972).

Although defined for complex numbers, the Fourier transform in data analysis is applied more often to real numbers. The identity

$$\exp(iy)=\cos(y)+i\sin(y),$$

which follows from a power series expansion, allows ϕ to be represented

$$\phi=(\phi_1,\phi_2) \tag{84}$$

for real z, where ϕ_1 and ϕ_2 are the cosine and sine transforms:

$$\phi_1=\sum_{t=0}^{N-1} z[t] \cos(2\pi tu/N)$$

$$\phi_2=\sum_{t=0}^{N-1} z[t] \sin(2\pi tu/N) \tag{85}$$

The obvious computation of the Fourier transform or its components is the direct application of (85). When N is small this is a reasonable practical procedure. The definition of small, as always, depends on the computing environment and the application, but a bound somewhere between 10 and 100 would be typical.

For larger series a more accurate and economical technique is the *fast Fourier transform* (FFT). The history of the FFT is open to different interpretations, but it became well-known after being discovered, rediscovered, or publicized in the paper by Cooley and Tukey (1965). Since then there have been extensive discussions and elaborations. Published algorithms in ALGOL60 are in Singleton (1968) and in FORTRAN are in Singleton (1969).

The essence of the FFT is reasonably simple. Let

$$w=\exp(i2\pi/N).$$

Then the transformation (82) can be written as a matrix multiplication

$$\phi=W\cdot z \tag{86}$$

where the matrix, W, has a very special form; namely, $W[j,k]=w^{jk}$, again indexing j and k from 0 for notational convenience. The essence of the algorithm is to show that, if N has an integer factor, say,

$$N=N_1\times n_1,$$

then the matrix, W, can also be factored:

$$W=P_1\cdot Q_1\cdot R_1 \tag{87}$$

In (87), P_1 represents a permutation of rows, Q_1 represents a set of n_1 separate Fourier transforms on subseries of length N_1, and R_1 has only n_1 nonzero elements in each row. Multiplying by R_1 then involves $O(Nn_1)$ operations and multiplying by Q_1 is equivalent to n_1 FFT's of length N_1. Therefore, letting c(N) be the estimated cost of the FFT on a series of length N,

$$c(N)=O(N\times n_1)+n_1\times c(N_1) \tag{88}$$

ignoring the permutations. If N_1 has factors we can repeat the process. Finally, if

$$N=n_1n_2...n_L \tag{89}$$

is a prime factorization of N, the cost of the FFT, by this argument, is

$$C = N\sum_{r=1}^{L} O(n_i). \tag{90}$$

When $N=2^L$ this gives rise to the common $C=O(N \log_2 N)$ statement. Direct application of the definition (76) requires $O(N^2)$ operations. The FFT is also much more accurate for large N than the direct calculation (Gentleman and Sande, 1966; Kaneko and Liu, 1970).

Some of the practical questions in using an FFT algorithm are as follows.

Real data. In data analysis the transform is generally applied to real data rather than complex, the purpose being to compute one or both of the series

$$\sum_{t=0}^{N}{}'' x[t] \cos(\pi tu/N)$$

$$\sum_{t=0}^{N}{}'' x[t] \sin(\pi tu/N) \tag{91}$$

The operation, $\sum''$, as in the discussion of integration in Section **g**, means that the first and last terms are halved. Notice the difference in range from (85).

When both transforms are wanted, one can take advantage of inherent symmetries to process a series of 2N reals as N complex numbers. If x_* is of length 2N, one can form a complex series z with

$$z[t]=x_*[2t] + i\, x_*[2t+1].$$

The standard complex FFT is applied to z and a post-processing step then disentangles the sine and cosine series. See Singleton (1968; 1969) for ALGOL60 and FORTRAN algorithms.

If one wants only one of the series above, a modification of the basic algorithm is needed. By using the even or odd behavior of cosine or sine, one of the series in (91) may be extended to the range of (85). The FFT logic can then be examined and modified to permit evaluation of the series in the minimal storage space and with about half the arithmetic for the previous case. See Gentleman (1972, 344-347).

Value of N; trigonometric function values. The derivation of the FFT makes clear that its cost (90) depends on the set of *prime factors* of N; namely, $2 \leqslant n_1 \leqslant n_2 ... \leqslant n_L$ such that n_j is a prime and

$$N = \prod_j n_j, \tag{98}$$

The most efficient programs will be specialized to certain classes of (98), such as $N=2^L$, $N=3^L$ or $N=4^L$, in the case of complex z.

Complications and increased costs accumulate as the largest prime in (98) increases. Temporary storage requirements are $O(n_L)$. The amount of arithmetic relative to a naive transform goes up as higher primes appear. In particular, if N is prime the FFT has no way to improve upon the naive method.

A related difficulty is that the most efficient mechanism for generating the trigonometric values in (93), for example, is to use the multiple angle formulas. Actual calls to library routines are required only for $\cos(\theta_1)$ and $\sin(\theta_1)$, where $\theta_1=2\pi/m$ for some m. Then the formulas for $j\theta$ are given iteratively:

$$\cos(j\theta_1)=\cos((j-1)\theta_1)\cos(\theta_1)-\sin((j-1)\theta_1)\sin(\theta_1)$$

$$\sin(j\theta_1)=\cos((j-1)\theta_1)\sin(\theta_1)+\sin((j-1)\theta_1)\cos(\theta_1)$$

The work involved, the storage required and the possible accumulation of rounding error increase with the prime factor involved.

For large N, one may then have to give an algorithm, even the general mixed-radix algorithm, a different value, N_*, with small prime factors. Singleton (1969) gives a useful table of "good" N values up to 100,000. The standard approach for a series of a bad length is to pad it with trailing zeroes to a good length, $N_*>N$. This has the effect that one computes ϕ_* with

$$\begin{aligned}\phi_*[u] &= \sum_t z[t]\exp(i2\pi tu/N_*)\\ &= \sum_t z[t]\exp(i2\pi tu_*/N)\\ &\text{"="}\ \phi[u_*]\end{aligned}$$

where u_* is a nonintegral value,

$$u_*=uN/N_*$$

Since one has merely rescaled the frequency domain, qualitative properties of the transform are preserved.

Power spectrum. The central statistical tool in assessing periodic behavior of time series is the real-valued function, s(u), which is the squared length of the Fourier function, $\phi(u)$, in (80) and is called the power spectrum:

$$s_F(u)=||\phi(u)||^2$$

The discrete Fourier series then generates the corresponding power spectrum:

$$s_F[u]=||\phi[u]||^2.$$

The *complex conjugate* of z=(a,b) is defined as conj(z) = (a,−b). Notice that

$$s_F = \phi \times \mathrm{conj}(\phi).$$

When the spectrum is to be calculated from raw data subject to error, some additional calculations are needed to ensure statistically useful estimates:

1. trends (linear or higher order) need to be removed;
2. the tendency of strong periodic components to leak into other periodic components should be reduced (often by *data windows* or *tapers*);
3. the spectral estimates should be smoothed, by a *spectral window*.

Some fairly standard choices exist for these operations; for example, *Hanning* and *moving averages* for points 2 and 3. See Jenkins and Watts (1968) and Bloomfield (1976). The latter contains some FORTRAN programs.

Convolutions; Variances and Covariances. An extremely important property of the Fourier transform is that it transforms convolutions into products. Specifically if z and z_* are two series we define their cyclic convolution at lag k by

$$v[k] = \sum_{t=0}^{N-1} z_*[t]\, z[(t+k) \bmod N].$$

Let ϕ and ϕ_* be the Fourier transforms (82) of z and z_*. Then the Fourier transform of v[k] is the product of ϕ and ϕ_*, and an alternative computation for v is

$$v[k] = \frac{1}{N} \sum_{u=0}^{N-1} \phi[u]\phi_*[u] \exp(-i2\pi uk/N).$$

See, for example, Bergland (1969). The computation of v then proceeds by three applications of the FFT: first to z and z_*; then to the product $\phi \times \phi_*$, to obtain v. For large N there will be very considerable savings over the direct formula. Because one does not usually want cyclic convolutions it is conventional to extend z and z_* by N added zero elements, in which case v reduces to the ordinary convolution,

$$\sum_{t=0}^{N-1-k} z_*[t]z[t+k]$$

but the FFT must be applied to 2N elements.

Two special cases of the convolution method are auto- and cross-covariance functions. The cross-covariance of z and z_* is

$$v[k]=\sum_{t=k}^{N-1} z[t]\,\mathrm{conj}(z_*[t-k])$$

and this becomes the more common auto-covariance when $z=z_*$. The convolution theorem immediately applies. The auto-covariance and cross-covariance are the inverse transforms of the spectrum and cross-spectrum, since the Fourier transform of the conjugate is the conjugate of the Fourier transform.

i. Parallel Computation

The analysis of efficiency for numerical methods presented in this chapter, and in Chapters 5 and 6, has been based on the concept of *serial* processing; that is, the user views computations as if they took place one at a time on scalar (integer or real) operands. The time taken for the total algorithm is then measured by the sum of the times for individual operations. Serial processing is the appropriate model for most computations done either on a dedicated machine with serial (single processor) hardware, or on a multi-user system where charges are applied more or less on the basis of total processor time used.

Other modes of computing are possible, however, and have received increasing attention, particularly in applications with real-time constraints (on-line monitoring, guidance, orbital calculations, etc.). The most general form would be a multiprocessor system running several independent programs in parallel. In fact, most large time-shared systems are of this form. Their use has not generated much new analysis of numerical methods, because users are not primarily charged for clock time elapsed, but instead for total resources used.

A simpler arrangement, which has generated considerable new analysis, is the use of *parallel computing,* in which one program has access to a number of processors. The easiest way to visualize parallel computation is to imagine the operators discussed in Section **b** applied to vector registers, so that RA and RB are now vectors of length p, say. Supposing that the number of cycles (i.e., the clock time elapsed) is the important quantity, parallel computation can profoundly alter the economics of many algorithms. A closely related concept is *pipeline computation,* in which instructions are loaded into a single register and then applied to a vector of operands. Analagous increases in speed are available in this case. For a general review of parallel and pipeline computation, see the March, 1977 issue of *Computing Surveys.*

Some operations are obvious candidates for parallel computation, but others must be radically redefined to obtain the full benefit. As an example of the former, consider the estimation of the partial derivatives of a function, $F(\theta)$, by simple differences (see Section **6.c**). We wish to compute the vector of values

$$g[j] = (F(\theta+\delta[j]e_{j,m}) - F(\theta))/\delta[j]$$

where θ is an m-vector, $e_{j,m}$ is the jth column of I_m and δ is a vector of step sizes. If we are doing serial computation, the arithmetic alone will require m steps of subtraction and division. In parallel computation, if there are at least m processors, only 1 cycle of each step is required. More interesting is the possibility of computing all the values of F in parallel. Notice that this suggests a programming language prepared to treat all arguments, scratch space, etc., as vectors of up to p values.

As another example, consider the computation of the inner product of two vectors, x and y, of length N:

$$x \cdot y = \sum_{i=1}^{N} x[i] \times y[i].$$

Suppose we have at least N processors. A parallel algorithm for $x \cdot y$ might be as follows:

I. Form RA = x·y; n = N

II. While n > 0 do
{ n = **ceil**(n/2)
(RB[i] = RA[2i] + RA[2i+1], i = 1,n);
RB[n+1 ,..., 2n] = 0; Interchange RA, RB}.

The function **ceil** is the smallest integer not less than its argument. The second step is done in a parallel loop. Notice that the upper half of the registers can be shut off after each step. It is easy to see that exactly k iterations of II are required, where k = **ceil**($\log_2 N$). By comparison, the serial computation of $x \cdot y$ takes on the order of 2N operations. The ratio

$$\log_2 N / N$$

represents the clock-time reduction in computation for a wide range of procedures, partly because the reduction by combining pairs of elements in II is clearly applicable in many calculations. The procedure is called *fan-in* because of its appearance as a tree with N results at the lowest level, N/2 at the next, and so forth. Savings in time are possible with fewer than N processors, as well. In this case, more data movement in and out of the registers is required. The number of numerical steps required, $m(p,N)$, may be computed from

$$m(p, N) = \min(k, \mathbf{ceil}(\log_2 N))) + \max(0, \mathbf{ceil}((N-2^k)/p))$$

where $k = \mathbf{ceil}(\log_2 p)$. For a general discussion, see Heller (1976).

From these simple examples, it can be seen that many operations in linear algebra, time series computations, integration, and other areas can be treated by parallel algorithms. On the other hand, a number of practical problems remain before easy, efficient use of parallel architecture is possible. Also, the computers providing this approach are not as yet either standardized or widely available. Sufficient work has been done, however, to demonstrate the relevance of the concepts in applications with strong emphasis on real-time efficiency.

Problems.

1. Given integers of L bits plus a sign bit, how could the logical operations (AND, OR, XOR, SHIFT, and ROTATE) be implemented using only arithmetic operations on integers? How would such a system be made portable among computers, first on the assumption that bit-strings can never be longer than L bits on the local machine, and then for bit-strings of general length?

2. (Blue, 1977) The simple implementation of inner product in Equation (13) of Section **c** ignores the possibility that a partial sum may cause overflow or underflow (i.e., be too large or small in absolute value to store as an intermediate result). For the special case of a vector norm

$$||x|| = (x \cdot x)^{1/2}$$

design an algorithm that avoids the possibility of either. (Hint: accumulate separately the three sums:

$$c_{BIG} = \sum_{|x_i|>B} (x_i/C)^2$$

$$c_{MIDDLE} = \sum_{B \geq |x_i| \geq b} x_i^2$$

$$c_{SMALL} = \sum_{|x_i| < b} (x_i/c)^2$$

where b and c are small constants, and B and C are large constants.) Could the idea be extended to general inner products?

3. Show how to use the normalized B-splines of (61) and (62) to generate a monotone approximation to a probability distribution, F(x), given the ability to calculate the density function, f(x) = dF(x)/dx. Show that with the knot positions, z_i, fixed the least-squares approximation is the solution to a constrained linear least-least squares problem.

CHAPTER FIVE

Linear Models

This chapter presents computational methods for the linear regression model and for related topics, including the analysis of variance and methods of multivariate statistical analysis (principal components, canonical analysis and discriminant analysis). For further background, readers may consult Rao (1973) or Gnanadesikan (1977) for the statistical and data analytic results. The relevant computational theory is chiefly that of *numerical linear algebra.* Stewart (1973) gives a general treatment of the computational aspects; Lawson and Hanson (1974) give a detailed discussion of least-squares.

a. Linear Regression.

The problem of linear regression is the fitting of a set of, say, N observations on one or more variables to a corresponding set of N observations on one or more other variables. Regarding the two sets of observations as the N by q matrix Y and the N by p matrix X, the problem is then to find a linear model, $X \cdot B$, for some set of coefficients, B, that approximates well to Y. The classical formulation is to minimize the sum of the squares of the residuals,

$$Z = Y - X \cdot B, \tag{1}$$

or in terms of matrix norms to minimize

$$\begin{aligned} S^2 &= ||Z||^2 \\ &= ||Y - X \cdot B||^2 . \end{aligned} \tag{2}$$

Since each column of (1) can be treated separately, there is little loss in specializing to the case of one Y variable, and the notation is much simpler. We assume $q = 1$ unless explicitly noted.

The classical probability model corresponding to (1), in its fullest form, assumes that for $B = B_T$, some generally unknown true value, the elements of Z are identically and independently distributed according to the normal distribution with expected value 0 and variance σ^2. Least-squares estimates of B then have several optimal properties, and the distributions of S^2, B and Z are known. (See Rao, 1973, Chapter 4, for example.) The statistical properties need to be formulated in an appropriate and natural way for computation. (See Section **d**.)

The essential ingredient in modern computations is the *orthogonal basis* or *orthogonal decomposition.* There exist a number of different procedures for computing such decompositions, but the differences among them are secondary. The key to effective computation is to have a flexible, reliable, general procedure. This should provide, easily, all the relevant summaries of the linear model. Although none of the existing published algorithms is entirely satisfactory, good procedures may be built upon algorithms for any of the orthogonal decompositions described in Section **c**. Throughout this chapter, the notation "·" stands for matrix product or vector inner product and the notation "′" stands for matrix transpose.

b. Orthogonal Bases.

The heart of numerical calculations for linear models is the concept of *orthogonality*.

Given an N by p matrix, X, we look for an N by r matrix, Q, forming a *basis* for X and having unit orthogonal columns. That is, every column of X is a linear combination of columns of Q; say,

$$X = Q \cdot R \tag{3}$$

for some r by p matrix R, and

$$Q' \cdot Q = I \tag{4}$$

where I is the r by r identity matrix. Further, if r is the *smallest* value for which we can find such a Q, r is the *rank* of X.

An orthogonal basis gives a solution to any linear least-squares regression on X. For, if Q is a basis for X, then for any linear combinations, X·B, of columns of X there is an r by q matrix, C, such that

$$X \cdot B = Q \cdot C. \tag{5}$$

Therefore (2) can be replaced by

$$S^2 = ||Y - Q \cdot C||^2, \tag{6}$$

and (6) can be minimized simply by

$$C = Q' \cdot Y. \tag{7}$$

This gives the least-squares values for S^2 and the residuals Z. If the actual coefficients, B, are desired, one must solve the equations

$$R \cdot B = C. \tag{8}$$

The computation of (8) depends on the form of the orthogonal decomposition. The common case is that R is an upper-triangular matrix and (8) is solved by back-substitution (see Section c).

For computation, however, it is better to think of C, R, and Q as the essential summary of the regression. These are less affected by numerical problems and can be used to generate all the relevant statistical information.

The proof that (7) is a least-squares solution is briefly as follows. The N by p matrix Q can always be extended to an N by N orthogonal matrix

$$Q_* = [Q\ Q_0]. \tag{8}$$

Note that $Q_*' \cdot X = R_*$, where R_* is an N by p matrix with R in the upper r rows and the remaining N - r rows all zero. Since Q_* is an orthogonal matrix

$$\begin{aligned} Y &= (Q_* \cdot Q_*') \cdot Y \\ &= Q \cdot Q' \cdot Y + Q_0 \cdot Q_0' \cdot Y . \end{aligned} \tag{10}$$

Substituting this into (6),

$$\begin{aligned} S^2 &= ||Y - Q \cdot C||^2 \\ &= ||Q_0 \cdot Q_0' \cdot Y + Q \cdot (Q' \cdot Y - C)||^2, \\ &= ||Q_0' \cdot Y||^2 + ||Q' \cdot Y - C||^2, \end{aligned} \tag{11}$$

the last line following because the columns of Q_* are unit orthogonal. The first term is independent of C, and the second is made zero by $C = Q' \cdot Y$, proving the result. (As a pedagogical aside, the derivation of (7) is simpler and more to the point than the discussion of "normal equations" still found in statistics texts.)

c. Orthogonal-Triangular Decompositions.

The most common form of orthogonal decomposition takes R to be upper-triangular

$$R[i,j] = 0 \quad \text{if } i > j. \tag{12}$$

The major alternative is the singular-value decomposition, discussed in Section **e**. The triangular form generally requires less computation and is more accurate, in the least-squares sense (see Section **g**). The singular-value form has important analytical advantages; for example, in cases where the rank of X is uncertain. It is also the preferred way to compute principal components and other multivariate summaries, as discussed in Section **k**.

There are three main ways to compute an orthogonal-triangular decomposition: elimination (the Gram-Schmidt method), reflection (the Householder transformation), and rotation (the Givens transformation). The relative merits of the three are discussed at the end of this Section and in Section **g**. The choice among the three methods is not particularly critical, however: the essential requirement is a numerically sound, flexible algorithm whose application to regression is well understood.

The Gram-Schmidt method iterates on X a column at a time producing at step j the jth column of Q, the jth row of R and (optionally) the jth element of C. Suppose we let $X_0 = X$. Step j of the method computes:

$$R[j,j]^2 = \sum_{i=1}^{N} (X_{j-1}[i,j])^2 ;$$

$$Q[i,j] = X_{j-1}[i,j]/R[j,j], \quad i = 1,...,N;$$

$$R[j,k] = \sum_{i=1}^{N} X_{j-1}[i,k]\,Q[i,j], \quad k > j;$$

$$C[j] = \sum_{i=1}^{N} Y[i]\,Q[i,j]. \tag{13}$$

Then the remaining columns of X_{j-1} are updated to form X_j by

$$X_j[i,k] = X_{j-1}[i,k] - Q[i,j]\,R[j,k] \tag{14}$$

for $i = 1,...,N$ and $k = j+1,...,p$. The actual implementation of (13) would usually overwrite X with X_1, X_2, and eventually Q (see Section **m**). The crucial computations in (13) are the *dot* or *inner* products in the first, third and fourth lines. These should be accumulated in double precision for best numerical results (see **g**), and can conveniently be written as a special routine as discussed in Section **4.c**. The dot product is sufficiently important that it may pay to fine-tune it for accuracy and speed.

The order in which the Gram-Schmidt calculations are done can have a considerable effect on the accuracy. The form in (13) and (14) is the so-called *modified* algorithm shown by Björck (1967) to be superior numerically. Other older forms, while equivalent algebraically, are potentially much less accurate and should not be used.

The theoretical error analysis and experience in practice confirm that the decomposition is competitively accurate in representing the regression by R and C. However, for some auxiliary results and applications it is important that Q be accurately orthonormal. If X is ill-conditioned, this property may be arbitrarily badly in error. The remedy is to reorthogonalize the matrix Q, when this property is important. See Section **f**, and Daniel et al. (1976) for a complete analysis and ALGOL60 algorithms.

The Householder and Givens methods operate somewhat differently in that a sequence of elementary reflections or rotations are computed to define the complete orthogonal transformation, Q_*. At the same time, the N by p matrix $R_* = Q_*' \cdot X$ is formed. While the complete transformation is defined, it will usually be available only in a specially coded form. Auxiliary algorithms should be provided to simulate multiplication by Q_* or Q_*'.

An elementary Householder reflection is defined by a vector, u, and operates on an arbitrary vector, x, to produce the vector $H \cdot x$, say, where

$$H \cdot x = x - 2(x \cdot u / u \cdot u) u. \tag{15}$$

Geometrically, $H \cdot x$ is a reflection in the plane perpendicular to u. The principal application of (15) is to transform a specific vector, w, into one whose elements are all zero, except the first. This can be done, letting e_1 be the vector $(1,0,0, \cdots ,0)$, by setting

$$u = w \pm ||w||e_1. \tag{16}$$

Direct substitution in (15) shows that $H \cdot w = \mp ||w||e_1$. The transformation (16) is said to *annihilate* the elements $w[2], \cdots$. The *Householder decomposition* produces (3) by applying p such transformations to the columns of X:

$$H_p \cdot H_{p-1} \cdots H_1 \cdot X = R_*$$

The jth transformation chooses u in (16) to be zero in the first $(j-1)$ elements, and chooses the remaining elements of u to annihilate the elements $j+1, \cdots, N$ in the jth column of $H_{j-1} \cdots H_1 \cdot X$. After p such steps, all the subdiagonal terms have been annihilated, as desired. Similarly, one may form $C_* = Q_*' \cdot Y$. It is possible and desirable to retain the full N elements of C_*, since these give an accurate representation of the residuals, as in (11).

While Householder reflections annihilate all the subdiagonal elements in a column at one time, elementary rotations annihilate a single element by rotating it with the diagonal. Only two rows are involved, say j and k. The transformation takes x into $w = G \cdot x$, where

$$\begin{aligned} w[j] &= c \times x[j] + s \times x[k] \\ w[k] &= -s \times x[j] + c \times x[k] \end{aligned} \tag{17}$$

and all other elements remain the same. To annihilate x[k], one takes

$$\begin{aligned} r^2 &= (x[j] + x[k])^2 \\ s &= x[k]/r;\ c = x[j]/r. \end{aligned} \tag{18}$$

To form the triangular decomposition one applies a sequence of such rotations to annihilate all the subdiagonal elements, in any order that does not reintroduce nonzeros. The most important such order works one *row* at a time. For the ith row, one takes $k = i$ and successively $j = 1, \cdots, \min(p, i-1)$.

Older methods, not based on orthogonal decompositions, are easily derived from the decomposition. Substituting (7) into (8) and multiplying on the left by R',

$$R' \cdot R \cdot B = R' \cdot Q' \cdot Y$$

But $R' \cdot R = X' \cdot X$ and $R' \cdot Q' = X'$, so that

$$(X' \cdot X) \cdot B = X' \cdot Y$$

These are, of course, the *normal equations* for B. Although drummed into the minds of generations of statistical students, they have limited practical relevance. The solution of the normal equations by Gaussian elimination or related methods had advantages for hand calculation (particularly when square roots were difficult). One related modern method is to compute R from the relation

$$R' \cdot R = X' \cdot X$$

by the *Choleski* decomposition. This is not generally recommended, because of potential inaccuracy (see Section **g**) and because it gives less information, in some respects, than orthogonal decompositions, although it requires less arithmetic than the latter methods. The method computes the rows of R successively. Suppose we have at hand R_j, the upper left-hand j by j submatrix of R. Then the first j elements of column j+1 of R, say r_{j+1}, are found from the same elements of $S = X' \cdot X$, say s_{j+1}, by back-solving

$$R_{j+1} \cdot R_j = s_{j+1} \tag{19}$$

and

$$R[j+1,j+1] = (S[j+1,j+1] - ||r_{j+1}||^2)^{1/2}$$

In assessing the relative merits of different orthogonal decompositions, the criteria of usefulness, reliability and convenience should be considered. The Gram-Schmidt method returns Q explicitly, making a number of auxiliary calculations easy to perform. It is essential, however, to use an accurate version of the algorithm. If Q is to be used subsequently, it should be accurately orthogonal, as in the algorithm of Daniel et al. (1976). The Householder method gives the simplest *basic* algorithm providing full accuracy and information. For some applications, the availability of the full transformation, Q_*, is important. Current implementations, however, are not as convenient as they should be, particularly in providing auxiliary algorithms to apply Q_* and Q_*'. Givens method is not in general as convenient or quite as accurate, but is preferred when access is to be by single rows or when general, sparse problems are encountered; i.e., when X has many zero elements, but not in a pattern such as the analysis of variance handles. Sections **g** and **m** provide further comparisons and recommendations.

d. Statistical Summaries.

The application of an orthogonal decomposition to regression is straightforward, but requires some auxiliary calculations not always included in basic numerical algorithms. As noted in Section **b** the essential summary of the regression consists of Q, R, and $C = Q' \cdot Y$. While various exotic schemes exist for packing these summaries into the original X and Y arrays, the simplest assumption is usually the best; namely, C is a vector of length p and R is a p by p matrix. The form of Q is determined by the algorithm, but in the Gram-Schmidt case it will also be the obvious: namely, an N by p matrix. Other schemes save an amount of storage (generally minor) but complicate auxiliary calculations, make examination of the summaries more difficult, and tend to confuse the user.

The regression coefficients in terms of X are given by the solution of $R \cdot B = C$. In the case that R is upper-triangular, this is the simple process of back-substitution: for $j = p, p\text{-}1, \cdots, 1$

$$B[j] = \left(C[j] - \sum_{k=j+1}^{p} R[j,k]B[j] \right) / R[j,j]. \tag{20}$$

Like the inner product, this basic back-substitution occurs so often that it is a candidate for a carefully written, high-precision special algorithm. Although (20) is numerically stable as a computation, it should be noted that B is more sensitive to numerical properties of X than the basic summary, (C,R,Q). It may be computed and studied but should not be used for *further* computation if possible.

The residuals, defined formally as

$$Z = Y - X \cdot B \tag{21}$$

are computed differently for different algorithms. The Gram-Schmidt method suggests the calculation

$$Z = Y - Q \cdot C \tag{22}$$

the actual calculations for (22) are the same as (14); therefore, the simplest method is to store Y in column $(p+1)$ of X, so that the same column will contain Z after X has been decomposed. The other two decompositions implicitly perform multiplications by Q_*', the full N by N orthogonal matrix, but do not form Q explicitly. One then needs some extra calculations to determine Z. Suppose

$$Q_*' \cdot Y = C_* = \begin{bmatrix} C \\ C_0 \end{bmatrix} \tag{23}$$

Then, from (22) and (10),

$$\begin{aligned} Y - Q \cdot C &= Q_0 \cdot C_0 \\ &= Q_* \cdot \begin{bmatrix} 0 \\ C_0 \end{bmatrix} \end{aligned} \tag{24}$$

The Householder decomposition allows us to multiply by Q_*, by applying the reflections H_j in the order $p, p-1, \cdots, 1$. If Y is stored as column $(p+1)$ of X, again, C_* will be computed by the decomposition. Unfortunately, most current published algorithms concentrate either on the basic decomposition or on B, and do not provide the auxiliary information in a convenient form.

Given the usual statistical model, one may estimate the *variance* matrices of the coefficients or the residuals. The variance, σ^2, of the errors in the standard probability model for (1) has the estimate

$$s^2 = ||Z||^2/(N-p). \tag{25}$$

Taking X, and hence Q, as fixed, the variance matrix of C is, assuming the errors to be independent and identically distributed

$$Q' \cdot \text{Variance}(Y) \cdot Q = \sigma^2 I.$$

Therefore, the elements of C are independent with variance estimated by s^2. The variance of B then follows from $R \cdot B = C$. Then S_B, the corresponding estimated variance matrix, can be computed from

$$R \cdot S_B \cdot R' = s^2 I \; ; \tag{26}$$

specifically,

$$S_B = s^2 R_+' \cdot R_+ \tag{27}$$

where R_+ is obtained by back-substitution in

$$R \cdot R_+ = I. \tag{28}$$

The variance matrix of the residuals comes directly from (22), writing Z as

$$\begin{aligned} Z &= Y - Q \cdot Q'Y \\ &= (I - Q \cdot Q')Y \end{aligned} \tag{29}$$

From (29), an estimate for the variance matrix is:

$$\begin{aligned} S_Z &= s^2 (I - Q \cdot Q') \cdot (I - Q \cdot Q')' \\ &= s^2 (I - Q \cdot Q'). \end{aligned} \tag{30}$$

this may be computed directly from the Gram-Schmidt results, and indirectly from the other decompositions.

These are the major summaries of regression models. The remainder of the section discusses some more specialized questions. Statisticians frequently remove the column means of X and Y before doing the regression, rather than including a column of 1's in X. One may then recover the decomposition that *would* have been obtained, had the initial column, **1**, been inserted. Let Q_I, R_I, and C_I be the summary for the latter case. Then if Y and X have had column means removed, the resulting Q is orthogonal to **1**, and

$$Q_I = [N^{-1/2}\mathbf{1} \quad Q]$$

$$C_I = [N^{1/2}\bar{y} \quad C]$$

$$R_I = \begin{bmatrix} N^{1/2} & r_I \\ 0 & R \end{bmatrix} \tag{31}$$

where r_I is the back-solution of

$$R \cdot r_I = N^{1/2}\bar{x} \tag{32}$$

and $\bar{x}$ and $\bar{y}$ are the column means removed from X and Y. One may then back-solve $R_I \cdot B_I = C_I$ to obtain coefficients. As to whether one *should* remove means, the answer is less clearcut. The need for scratch space may be reduced, since the regression may be done in place by overwriting X. Numerically, one can show that, provided the column means are large relative to the column standard errors, the reduced matrix is better conditioned, in the sense of Section **f**. In spite of remarks that I and others have made, however, this does not automatically imply that the least squares calculations will be more accurate; this seems to be an open question.

Two N by N matrices are of interest. From (29) the matrix, $(I - Q \cdot Q')$, represents the *projection matrix* onto the space of residuals from X; that is, this matrix, multiplied into any vector, Y, gives the residuals from fitting Y to X. Similarly, the matrix

$$H = Q \cdot Q'$$

projects Y onto the fitted values, say $\hat{Y}$. (This matrix is sometimes called the "hat" matrix.) One would not normally form such large matrices, and fitted values and residuals may be computed directly as noted. However, the ith diagonal element of H has been proposed as a measure of the effect of the ith observation in regressing against X (the term *leverage* is used to describe this quantity). Equation (30) shows that the diagonal element of $I - H$ gives the variance of the ith residual. It may be computed as $1 - \sum_j Q[i,j]^2$, if Q is explicitly available. See Hoaglin and Welsch (1977) for further discussion.

A number of articles and books have discussed the notion of *generalized inverses* or *pseudoinverses,* which are related to the idea of representing the regression coefficients as

$$B = X_{+} \cdot Y \tag{33}$$

for some matrix, X_{+}. The usefulness of this approach in practical data analysis is doubtful, since the information it provides can be computed, more accurately, from the decomposition. However, X_{+} can be computed to satisfy (33) by back-solving the N sets of equations

$$R \cdot X_{+} = Q'. \tag{34}$$

In the case of doubtful or deficient rank, a similar calculation based on singular-values would be better defined.

e. Singular-Value Decomposition; Eigenvalue Decomposition

An extremely useful tool for studying linear models and a number of related problems is the following. Given any N by p matrix, X, there exists an N by N orthogonal matrix, U_*, a p by p orthogonal matrix, V_*, and an N by p diagonal matrix, D_*, with diagonal elements

$$d_1 \geqslant d_2 \cdots \geqslant d_s \geqslant 0 \tag{35}$$

where $s = \min(N,p)$, such that

$$X = U_* \cdot D_* \cdot V_*' \tag{36}$$

In practice, one generally needs only U, V, and D, being, respectively, the first s columns of U_* and V_* and the upper left s by s submatrix of D_*:

$$X = U \cdot D \cdot V' . \tag{37}$$

Besides providing the best approach to computation of principal components, canonical correlations, and discriminant analysis (see Section **k**), the decomposition (37) has a number of significant mathematical properties. The key to most of these is the following: let the function f(v) be defined for all p-vectors v such that $||v|| = 1$, by

$$f(v) = ||X \cdot v|| \tag{38}$$

Then the maximum of f(v) is d_1, attained when $v = v_1$, the first column of V, and the first column of u is $u_1 = X \cdot v_1 / d_1$. In fact this result can be used to derive (37). (See the addendum to this section.) Further, among all v orthogonal to the first j−1 columns of V, the maximum of (38) is d_j, attained at v_j, and $u_j = X \cdot v_j / d_j$.

The computation of the decomposition proceeds in two stages. First, the matrix X is reduced to *upper bidiagonal* form; i.e., such that only the [j,j] and [j,j+1] elements are nonzero. This is usually done by Householder reflections from the left and the right. Then an iterative method computes the decomposition of the bidiagonal form, and transforms back to obtain (37). The details of computation are important, but fortunately well understood. See, for example, Chapter 18 of Lawson and Hanson (1974). Algorithms are given in FORTRAN by Businger and Golub (1969) and Lawson and Hanson (1974, 256-263; 292-300), and in ALGOL60 by Golub and Reinsch (1970).

Returning to linear regression, we see that (37) gives an orthogonal decomposition as in (3) with

$$Q = U$$

$$R = D \cdot V'$$

Therefore a solution for C and B is

$$C = U' \cdot Y$$

$$B = V \cdot D_+ \cdot C \tag{39}$$

where D_+ is a diagonal matrix with diagonal elements $1/d_j$, if $d_j > 0$, and 0 otherwise.

Notice that (39) applies whatever the number of nonzero d_j. The importance of this result is in the case of uncertain rank for X. For applications where rank is not uncertain, singular values give a more expensive and somewhat less accurate solution to the least-squares problem (see Section **g**).

The singular-value decomposition is closely related to the *eigenvalue* decomposition. Let

$$S = X' \cdot X.$$

then from (37)

$$S = V \cdot \Lambda \cdot V' \tag{40}$$

where V is as in (37) and Λ is a diagonal matrix with diagonal elements, λ_j,

$$\lambda_j = d_j^2$$

Equation (40), the eigenvalue decomposition of S, provides older, alternative computations for principal components and related quantities, in much the same sense that the Choleski decomposition of S provides an alternative to orthogonal decompositions in least-squares regression. In both cases, computations based on orthogonal decompositions of X are more general and frequently more accurate.

There are several generalizations of (40). First, S may be any symmetric, real matrix. The λ_j will be positive (or nonzero) if and only if S is positive definite (or positive semidefinite), but the computations required are the same, and are closely related to singular-value calculations. When S is a general matrix the problem to be solved is stated

$$S \cdot V = V \cdot \Lambda$$

In this case the λ_j may be complex numbers and V will not generally be orthogonal. A different generalization is to the problem

$$S_1 \cdot V = V \cdot \Lambda \cdot S_2$$

with various assumptions about S_1 and S_2. This problem has statistical applications if S_1 and S_2 are positive semidefinite. Algorithms for eigenvalue problems are available in the excellent EISPACK collection prepared by the Argonne National Laboratories and implemented on several widely used computers. Published algorithms for the symmetric case are found in Businger (1965) and Bowdler et al. (1968) and for the general case in Martin and Wilkinson (1968). An algorithm for the general two-matrix case is described by Moler and Stewart (1973), and a FORTRAN program is in Kaufman (1975).

There are several connections between singular values and the notion of the *rank* of a matrix. In non-numerical contexts the rank is just the maximum number of columns or rows of X that form a set of *linearly independent* vectors. Vectors $x_1, \ldots, x_r$ are linearly independent mathematically if there is no nontrivial linear combination which is zero in all elements; i.e., if the equation

$$\sum_j a_j x_j = 0, \tag{41}$$

has only the solution $a_1 = a_2 = \cdots = 0$. However, this definition is not numerically or statistically meaningful. A first requirement for a computational definition is that the a_j be normalized; e.g.,

$\sum_j a_j^2 = 1$. Second, the rank should be invariant to multiplying all the values of X by a nonzero constant. Third, *exact* zero is not a reasonable test; instead, one should test for *near* zero, to some chosen tolerance, ϵ. (From the discussion in Section **4.b**, for example, it is clear that ϵ should be somewhat larger than the relative precision of the numbers in the calculation.)

A definition satisfying the requirements is the following: the *numerical* rank of X is r to a tolerance ϵ if there are exactly r columns (or rows) of X, say $x_1, \ldots, x_r$, such that

$$\min_a \sum_j a_j x_j \geqslant \epsilon \max_a \sum_j a_j x_j ,$$

for all normalized a_j.

This definition is equivalent to: the rank of X is r to a tolerance ϵ if

$$d_r/d_1 \geqslant \epsilon > d_{r+1}/d_1 \tag{42}$$

where d_j are the singular values in (37) and we define $d_{s+1} = \cdots = 0$. We now have a general and operational definition of rank, satisfactory for numerical work. The ratio used in this test is called the *condition number* of X, when given in the form

$$\kappa(X) = d_1/d_r$$

The following section pursues the questions of numerical and statistical rank further.

There are many other interesting properties of singular values, some potentially useful in a variety of applications. See Stewart (1973) and Golub and Styan (1973).

Addendum: Brief Derivation of (37)

Consider f(v) in (38). The set of unit v is closed, and f is finite and continuous on this set. Therefore, it attains a maximum, d_1, at some $v = v_1$. Similarly, among v orthogonal to v_1 there is a maximum, d_2, at v_2. Proceeding in this way we generate $d_1,...,d_s$ satisfying (35), and unit orthogonal $v_1,...,v_s$ to form V. Define $u_j = X \cdot v_j/d_j$ and $U = [u_1, \ldots, u_s]$.

The key point is then to prove that the u_j are orthogonal (clearly $||u_j|| = 1$). Suppose $u = X \cdot v$ is orthogonal to u_1. Then v is orthogonal to v_1, for otherwise we could write

$$v_1 = v_* + av$$

for some v_* orthogonal to v. This leads to a contradiction. Since

$$1 = ||v_1||^2 = ||v_*||^2 + ||a \cdot v||^2,$$

either $a = 0$ or $||v_*|| < 1$. Also,

$$X \cdot v_1 = d_1 u_1 = X \cdot v_* + aX \cdot v$$

and $X \cdot v$ was assumed orthogonal to u_1. Hence

$$\begin{aligned} ||X \cdot v_*||^2 &= ||d_1 u_1 - aX \cdot u||^2 \\ &= d_1^2 + a\,||X \cdot v||^2 \\ &\geqslant d_1^2, \end{aligned}$$

and if $a \neq 0$, $f(v_*/||v_*||) > d_1$, which is impossible. Thus the u_j must be orthogonal.

f. Condition; Rank; Iterative Improvement

The standard orthogonal decomposition of Section **c** gives a thoroughly adequate numerical solution for most linear least-squares problems. A smaller set of problems require special methods. Although there is a tendency to lump together problems of "deficient rank", there are actually at least three distinct cases. First, it may be known that X has by construction rank $r < p$ (as in the analysis of variance). Second, there may be reason to regard the rank as ill-defined, perhaps because of statistical uncertainty about the elements of X. Third, it may be necessary to guarantee high accuracy, for example, in the computed residuals, even for numerically difficult problems.

For the first case, one expects that the numerical rank of the matrix will usually be well defined. A general and readily interpretable test can be based upon the singular values, using a criterion such as (42), with ϵ determined by the precision of the calculations. When X has been constructed of rank r, a sharp break between d_r and d_{r+1} is usually found. If no such break occurs, the rank must be regarded as doubtful. In this case, the solution to the linear least-squares fit may also be ill-defined. Specifically, suppose the rank could be either r or r+1 for reasonable values of ϵ. Then the component of the linear fit represented by regressing Y on u_{r+1} is uncertain. If Y and u_{r+1} are uncorrelated, the prediction of Y is not harmed (although the coefficients are uncertain). If the correlation is high, however, the entire fit is in doubt.

The orthogonal-triangular decomposition can be used to provide a similar test for rank. This method does not give the general interpretation which singular values allow. However, it does appear to work well on most examples of the first class of problem, it is somewhat cheaper than computing singular values, and it leads directly to selecting a linearly independent subset of the original columns of X. By permuting the columns of X, one may arrange that the diagonal elements of R are monotone decreasing; i.e., there is a permutation, P, such that

$$X \cdot P = Q \cdot R$$

with R upper triangular and

$$R[1,1] \geqslant R[2,2] \cdots \geqslant R[p,p] \geqslant 0$$

For the Gram-Schmidt and Householder methods, working one column at a time, the algorithm looks at each stage for the largest column norm among the remaining columns, and pivots that column into the decomposition next. For these algorithms, little extra work is required over computing the decomposition without permutations. See Daniel at al. (1976) and Lawson and Hanson (1974) for further details and algorithms. The Givens method requires some extra work, after all rows have been processed (see Gentleman, 1974). The diagonal elements, R[j,j], may then be used for a test of numerical rank analagously to the use of the singular values. Again, when X has been constructed of rank r, we expect a sharp break after R[r,r]. Golub, Klema and Stewart (1977) give a detailed discussion. They construct a test (based on a somewhat different definition of numerical rank) which selects a well-defined linearly independent set of columns of X. The test is conservative, in the sense that the singular value approach may detect well-defined subsets which this test misses. Nevertheless, a test based on the orthogonal-triangular decomposition may be preferred for the reasons mentioned previously.

Now consider the second case: uncertainty in the elements of X. In this case, similar computations may be used, but the interpretation of the test must be different. Suppose the element X[i,j] differs from its "true" value by, say, e[i,j]. Then, *if* one may assume that the errors in the jth column of X are independently and identically distributed, and *if* one can estimate the standard deviation, σ_j, of that distribution, then the jth column of X should be divided by the estimate of σ_j. The rescaled elements of X now have errors with variance 1. It follows that unit linear combinations of columns of X, as in (41), have the same variance. Thus if

$$d_r >> 1 >> d_{r+1}$$

the singular values give a directly interpretable indication that the data has *statistical* rank r. The diagonal elements of the triangular decomposition may again be used for a similar test. The relative merits of the tests are much the same as in the first case.

The case of uncertain data is intrinsically less straightforward than the first case. The simplifying assumptions made to allow rescaling may not hold. It is necessary to estimate the error variance, σ_j, by some outside knowledge or observation. Even if a suitable scaling is obtained, it is not certain that the statistical rank of the data will be well defined. However, there is no justification for ignoring uncertainty in the data, particularly in applications to economics, sociology and other areas where such uncertainty is recognized to be substantial. Methods such as the one given here can then be of significant assistance. For example, the author applied this technique to Longley's (1967) data with plausible choices of σ_j, and found that his data had effectively $r = 3$ with $p = 6$ (Chambers, 1972). See also Problem 1 of this Chapter and Golub et al. (1977).

The third case takes us back to a more straightforward problem: we wish to ensure high accuracy in the least-squares sense for the coefficients and/or residuals. The technique of *iterative improvement* has been developed for this purpose. The coefficients B and residuals Z jointly satisfy the system of N + p linear equations:

$$\begin{bmatrix} I & X \\ X' & 0 \end{bmatrix} \cdot \begin{bmatrix} Z \\ B \end{bmatrix} = \begin{bmatrix} Y \\ 0 \end{bmatrix} \tag{43}$$

The first N equations define Z; the last p express the orthogonality of Z to X. Iterative improvement repeatedly corrects both Z and B to attempt to make the difference between the two sides of (43) tend to zero. Detailed analyses and an AL60L60 procedure are given in Björck (1967a, 1968).

The basic theoretical result is that the iteration should converge to give Z and B to essentially the relative accuracy of the arithmetic unless either X is so ill-conditioned the answers are undefined or the true values in (43) are essentially zero.

R. A. Becker and the author experimented with a FORTRAN version of Björck's algorithm. As theory predicts, the accuracy, measured by the maximum correlation between Z and X, remained of the order of ϵ (here between 10^{-8} and 10^{-9}). There was, however, a further result that prevented us from endorsing the method fully. When ordinary Gram-Schmidt was reprogrammed entirely in double precision, it obtained results at least as accurate as iterative improvement at significantly less cost (measured by processor time). One may therefore question whether iteration is an economical method for very high accuracy in least-squares. (The moral may be that a larger word length is sometimes as effective as a better algorithm.)

g. Cost and Accuracy

The three orthogonal-triangular decompositions do not differ greatly in cost or accuracy, and the typical user should give less weight to these questions than to convenience and appropriateness for the specific environment. The Choleski method is inferior in terms of accuracy for difficult problems. The singular value decomposition requires significantly more calculation and tends to provide somewhat less accurate results, in the narrow sense of solving the least-squares problem (Chambers, 1974). However, it is the recommended method when the more complete information about the structure of X is useful.

The cost of the numerical calculations for linear least-squares is moderately easy to estimate by counts of basic arithmetic operations. As discussed in Chapter 2 in general terms, operation counts are a reasonable approximation but not exact, especially for small amounts of data. Also, for the majority of users the amount of numerical calculation in linear regression is essentially trivial. On a large, commercial-type system, the user frequently pays more to have the data read and the results printed than for the calculations themselves. The amount of discussion in the literature concerning cost-reducing methods for these problems does not imply that cost should be the first consideration, and we have no hesitation in relegating it to a secondary role in our recommendations.

Consider the Gram-Schmidt calculations on X in (13) and (14). At stage j, there occur $p-j+1$ inner products of vectors of length N, 1 square root, plus $p-j$ divisions and $(p-j)N$ each of multiplications and subtractions. This gives an estimate of about Np^2 each of multiplications and additions, ignoring terms of lower order in N and p. The step of the back-substitution operation (20), by similar counting, takes about $\frac{1}{2}p^2$ additions and multiplications. If we go on to compute, for example, the variance matrix of the estimates as well, 2p of these back-substitutions will be required. Thus the overall operation count can be estimated in the form

$$aNp^2 + bp^3$$

for values of a and b in the range of small integers. This functional form is the most important cost result. It establishes least-squares as a "cubic" problem. The form applies to all five methods mentioned, with appropriate values of a and b. In increasing order of operation counts the methods may be ranked: Choleski, Householder, Gram-Schmidt, Givens and singular-value.

Accuracy will generally be more important than numerical cost to most users, although again one must remember that the available estimates are either bounds or approximations. Small differences in these may not imply any practical advantages. Most of the detailed error analyses are rather lengthy, but we may suggest here the flavor of the results.

Consider the Gram-Schmidt algorithm. Björck (1967) gives two basic bounds. First a perturbation result estimates the stability of the coefficients B and residuals, Z. Suppose X and Y are perturbed by δX and δY. The resulting coefficients and residuals are then perturbed by δB and δZ, usually involving the condition number (section **f**) which we denote, if X has rank r,

$$K = \kappa(X) = d_1/d_r \tag{44}$$

In particular, assuming $||B|| \neq 0$

$$\frac{||\delta B||}{||B||} \leqslant cK\left(1 + \frac{cK||Z||}{||X||\ ||B||}\right) + \frac{cK||\delta Y||}{||X||\ ||B||}$$

with c generally close to 1. The important qualitative implication is that, *if* the fit is good, so that

$$||Z|| << ||X||\ ||B||$$

then the sensitivity of B is essentially proportional to the condition of X. In contrast, any method of defining B from $X' \cdot X$ must depend on K^2, since the eigenvalues of $X' \cdot X$ are $d_1^2, \ldots, d_p^2$ and its condition number is $(\kappa(X))^2$. It is in this sense that orthogonal decompositions are inherently more accurate than Choleski methods.

Bounds on computational error are also available in the form of *backward* error bounds. These state that the computed decomposition is the exact decomposition of a matrix X + E and then bound the size of E. The bounds obtained depend upon whether double precision accumulation (see Chapter 4) is used for inner products. If so, one obtains the bound

$$||E|| \leqslant c(p-1)\epsilon||X|| \tag{45}$$

with ϵ the relative precision of the arithmetic, and c around 1. If double precision accumulation is *not* used the bound will be proportional to $N(p-1)\epsilon||X||$.

As noted in Section **c**, further calculations are required if the matrix Q is required to be accurately orthonormal. See Daniel et al (1976).

Similar results apply to Householder decompositions (Wilkinson, 1965, Chapter 3; Lawson and Hanson, 1974, Chapter 16) and Givens decompositions (Wilkinson, 1965; Gentleman, 1973). The former has slightly better error bounds than Gram-Schmidt, the latter slightly worse. The chief advantage of the Householder algorithm is that it provides a way to apply the full orthogonal transformation, Q_*, with high accuracy. No extra work is required in the decomposition to achieve this, but some auxiliary routines to apply the transformation are needed.

The singular-value method is not directly comparable. First, it depends on the scale of the columns of X more than the other methods. Second, the major rationales mentioned for its use imply a concern for other questions, such as uncertain rank, rather than a spirit of "least-squares regardless." It gives inherently more complete and appropriate answers, but in pure least-squares problems it is somewhat more expensive and less accurate (Chambers, 1974).

h. Weighted Least-Squares

The computational methods described extend to the case of weighted linear least-squares regression with known weights. With residuals defined as $Z = Y - X \cdot B$, the problem is now to minimize

$$\sum_{i=1}^{N} w_i Z[i]^2 = Z' \cdot W \cdot Z \tag{46}$$

assuming $w_i \geqslant 0$ and W the diagonal matrix with diagonal elements $w_1, \ldots, w_N$. The simplest solution to (46) is usually to form

$$X_w = W^{1/2} \cdot X; \quad Y_w = W^{1/2} \cdot Y; \tag{47}$$

i.e., multiply each row by the square root of the weight. Since

$$Z' \cdot W \cdot Z = (Y - X \cdot B)' \cdot W \cdot (Y - X \cdot B)$$
$$= (W^{1/2} \cdot (Y - X \cdot B))' \cdot (W^{1/2} \cdot (Y - X \cdot B))$$

the weighted regression has the same solution as the *unweighted* regression to minimize

$$||Y_w - X_w \cdot B||^2, \tag{48}$$

and no special algorithms are required.

A few details need to be kept clear. If one wishes to remove column means beforehand, these must be weighted means; e.g.,

$$\bar{y}_w = \sum_i w_i Y[i] / \sum_i w_{i.}$$

Note that this is *not* the unweighted mean of Y_w; the correct formula for getting Y_w from Y while removing the mean is

$$Y_w[i] = (w_i)^{1/2}(Y[i] - \bar{y}_w).$$

The process of putting the intercept term back in, as in (31), is modified. Let $N_w = \sum_i w_i$. Then

$$Q_I = [N_w^{-1/2} \, w^{1/2} \quad Q]$$
$$C_I = [N_w^{1/2} \, \bar{y}_w \quad C]$$
$$R_I = \begin{bmatrix} N_w^{1/2} & r_I \\ 0 & R \end{bmatrix}$$

where w is the vector of N weights and r_I is the found by back-substitution in

$$R \cdot r_I = N_w^{1/2} \, \bar{x}_w$$

All these results can be proved by imagining the first step of a Gram-Schmidt method to remove the column of weights, $w^{1/2}$.

It is necessary to distinguish two sets of residuals: the residuals, Z, on the original scale and

$$Z_w = Y_w - X_w \cdot B,$$

the residuals that come from regression of Y_w on X_w. Provided all the weights are positive, Z can be computed as

$$Z = W^{-1/2} \cdot Z_w$$

If some of the weights are zero, the corresponding residuals must be computed directly from $Y - X \cdot B$. Note that classical least-squares theory applies to Z_w, not to Z; e.g., the elements of Z need not be nearly identical in distribution. The variance matrix of Z is

$$S_Z = s^2 W^{-1/2}(I - Q \cdot Q')W^{-1/2}$$

from (22).

A second approach is in principle slightly more accurate, although slightly less convenient. Here one defines a *weighted* orthogonal decomposition

$$X = Q_w \cdot R_w$$

such that

$$Q_w' \cdot W \cdot Q_w = I.$$

If $X_w = Q \cdot R$ in the previous approach,
$R_w = R$.

However, Q_w and R_w come directly from a Gram-Schmidt decomposition in which all inner products are replaced by *weighted* inner products:

$$x \cdot y = \cdot \sum_{i=1}^{N} w_i x[i] y[i].$$

This is an easy modification to make in practice if inner products are a separate routine. Similarly C in (7) is computed by weighted inner products. Then B can be found by ordinary back-substitution. For further discussion see Chambers (1975).

i. Updating Regression

Given the solution to a regression problem one may wish to add and/or delete either observations or independent variables. Methods exist for each of these problems that reduce the amount of computation and, in the case of observations, the storage requirements over complete recomputation of the regression. Such savings will be relevant, *provided* the cost of the numerical calculations is itself relevant. There may also be advantages of convenience in not requiring all the data to be present at one time.

Updating for observations is the more common case. In

general, this can be formulated most easily in terms of weighted regression. Let

$$X = \begin{bmatrix} X_{11} \\ X_{21} \end{bmatrix}; \quad Y = \begin{bmatrix} Y_{11} \\ Y_{21} \end{bmatrix};$$

$$W = \begin{bmatrix} W_1 & 0 \\ 0 & W_2 \end{bmatrix}$$

represent $N = N_1 + N_2$ observations, and suppose the regression of Y_{11} on X_{11} has been computed. Letting W_2 have a weight $-w_i$ when we want to *delete* row i, any combination of additions and deletions corresponds to adding X_{21} and Y_{21} with weights W_2. The case of addition and/or deletion in ordinary least-squares corresponds to $w_i = \pm 1$. Any orthogonal decomposition can be applied to the problem, based on this representation. The most obvious and most convenient choice is the Givens method since it already works one row at a time. Chambers (1971) provides FORTRAN algorithms for the case of simple addition or deletion. Gentleman (1973; 1974) gives more complete procedures in ALGOL60 for general weights.

As the use of negative weights suggests, *deletion* of observations has none of the numerical guarantees of ordinary regression. Basically, the recomputed regression cannot be completely accurate if the data deleted dominates the remaining data. This will be true regardless of the computational method and has inherently nothing to do with ill-conditioning. Extreme cases of this loss of accuracy are easy to concoct but rare in practice (see Chambers, 1975). In long sequences of updating, however, it is good practice to recompute from scratch when the total number of observations handled has become much larger than the basic size of the regression, say by a factor of 10 or more.

The other decompositions can also be applied to the updating problem. The Gram-Schmidt method is described in Chambers (1975), the Householder in Golub and Styan (1973) or Lawson and Hanson (1974), and the singular-value method in Businger (1970). Except for specialized problems, there seems little reason to prefer these.

To achieve full numerical stability in updating, one must retain Q, the orthogonalized form of X. This destroys some of the cost advantages, but may be preferrable when accuracy is essential. Daniel et al. (1976) provide a full set of algorithms for updating a Gram-Schmidt decomposition, using Givens rotations for observations.

j. Regression by Other Criteria; Robust Regression

Although least-squares regression has a number of optimality properties, at least when the residuals are assumed to have independent, identical normal distributions, there is much interest in more robust criteria for fitting linear models. By *robust* one usually means that a technique has nearly optimal properties under the standard assumptions and remains a good method when various failures of the assumptions occur. A number of different approaches may be taken to this problem. For example, one may replace the sum-of-squares criterion by the general requirement to minimize

$$\sum_{i=1}^{N} \rho(Z[i]). \tag{49}$$

The function, $\rho(x)$, should generally be similar to x^2 for $|x|$ small, but increase less rapidly than x^2 when $|x|$ is large. Closely related to (49) is the solution of the equations

$$\sum w_i Z[i]X[i,j] = 0 \tag{50}$$

for $j = 1, \cdots, p$. The weights w_i are *not* constant, but chosen to give less weight to large residuals. Generally w_i is defined as some scaled function of Z[i], often in terms of an *influence function*, ψ:

$$w_i = \frac{\psi(Z[i]/s_Z)}{Z[i]/s_Z}$$

where $\psi(x)$ is like x near $x = 0$ but levels off or redescends to zero as $|x| \to \infty$. The scalar s_Z is some (robust) scale estimate of Z, such as the median absolute value. See Andrews (1974), Huber (1973), Hampel (1973) and the papers by Tukey and by Andrews in Lide and Paul (1974) for further discussion.

Equations (49) and (50) suggest the two main computational methods. One may treat the problem as one of *nonlinear* optimization and apply the techniques of Chapter 6. Alternatively, noting that (50) is a standard weighted least-squares procedure when the w_i are given, one may look for a *fixed point* of (50), by setting initial weights, computing the corresponding coefficients and residuals, forming new weights and repeating.

Iterating linear least-squares is simple and will usually work well if the linear model is reasonably appropriate. If there are serious outliers in the data, the method may perform poorly because the ordinary least-squares solution is too far from the desired robust estimate. A better starting estimate is obtained by minimizing the absolute residuals (the L_1 norm):

$$\sum_i |Z_i|$$

Algorithms for L_1 fitting are given in Barrodale and Roberts (1974) and Bartels and Conn(1978).

If numerical problems arise (such as nonunique solutions for different initial weights), one may wish to apply optimization directly. See also the general comments on fixed point methods in Section **6.h**. The use of a scale for the residuals, s_Z, which is not constant somewhat complicates the relation between (49) and (50). In practice, one may choose to keep the scale constant for several steps of iteration. With constant scale, (49) represents a convex function if ρ is convex, and convergence of (50) can be proved. While non-constant s_Z can be worked in without disturbing these results, it is common practice to use non-convex ρ, in particular, those corresponding to ψ which re-descend to zero for large argument. The convergence properties in this case, and the overall performance of the methods, need further study.

k. Principal Components; Canonical Analysis; Discriminant Analysis

The traditional treatment of these procedures in statistics texts is concerned with cross-product matrices and eigenvalue analysis. For modern discussions, these may usually be replaced by definitions in terms of data matrices and singular-value analysis. The result is a more elegant, general, accurate, and simple treatment that leads directly to computation. In this section we assume the reader to be acquainted with the statistical techniques. These are described in most texts on multivariate analysis (e.g., Dempster, 1969; Rao 1973; Gnanadesikan, 1977). We also assume, unless otherwise noted, that data matrices Y and X have column means removed.

Principal component analysis asks for the unit linear combination of columns of Y with the largest sample variance. If

$$Y = U_Y \cdot D_Y \cdot V_Y'$$

is the singular-value decomposition of Y as in (35) to (38) of Section **e**, it is immediately obvious that the linear combination desired is given by v_{Y1}, the first column of V_Y, that its variance is $d_{Y1}^2/(N-1)$, and the N data values are $d_{Y1}u_{Y1}$, the (scaled) first column of U_Y. Similarly the rest of the decomposition defines the remaining principal components. Thus the computations required are to remove column means and apply a singular value algorithm.

Canonical analysis may be defined in several ways, one of which asks for the linear combination, $y = Y \cdot a$, of columns of Y that is best predicted by regression on X, in the sense that the ratio

$$||y - X \cdot b|| / ||y|| \tag{51}$$

is minimized. Equivalently one may ask to maximize the ratio

$$||X \cdot b|| / ||Y \cdot a|| \tag{52}$$

for all a and b.

The computation proceeds as follows. Let

$$\begin{aligned} Y &= Q_Y \cdot R_Y \\ X &= Q_X \cdot R_X \end{aligned} \tag{53}$$

be orthogonal decompositions for Y and X. Define the p by q matrix

$$W = Q_X' \cdot Q_Y \tag{54}$$

Then the singular value decomposition of W defines the canonical analysis. If

$$W = U_W \cdot D_W \cdot V_W' , \tag{55}$$

the canonical correlations are the singular values, and the corresponding linear combinations of Y and X, say A and B, are given by back-solving the equations

$$\begin{aligned} R_Y \cdot A &= V_W \\ R_X \cdot B &= U_W \end{aligned} \tag{56}$$

The proof of these assertions begins by writing (52) as, say $||Q_X \cdot u|| / ||Q_Y \cdot v||$. From (29), the ratio can be shown to be equivalent to

$$||Q_X \cdot Q_X' \cdot Q_Y \cdot v|| / ||Q_Y \cdot v|| = ||Q_X' \cdot Q_Y \cdot v|| / ||v||,$$

because Q_X and Q_Y have orthogonal columns. The definition in terms of singular values then follows from the property (38) in

Section **e**.

A specific computational method would be as follows: compute Q_X and Q_Y from a Gram-Schmidt procedure, form W by inner products of the columns of Q_X and Q_Y, compute the singular-value decomposition and use a back-substitution routine for (56). Note that reorthogonalization of Q_X and Q_Y would be desirable for numerical accuracy (see Section **c**).

Multiple discriminant analysis can be regarded as the special case of canonical analysis in which X is a set of dummy variates, the ith of which is 1 for all observations in group i and 0 elsewhere. It can be seen in this case that R_X is a diagonal matrix with diagonal elements, $(n_1^{1/2}, \cdots, n_q^{1/2})$, where n_i is the number of observations in group i. Also, W is given by the matrix of group means:

$$W[i,j] = n_i^{1/2}\, \bar{y}_{ij}$$

where $\bar{y}_{ij}$ is the mean of the jth column of Y for the ith group. These two simplifications may be applied directly to the general scheme above for computing canonical analysis. At present, generally available algorithms implementing the procedures for canonical and discriminant analysis in a satisfactory way do not exist.

1. Analysis of Variance

We assume again that the reader is familiar with the basic statistical techniques (e.g., Scheffe, 1959). Suppose that the analysis of variance model is defined in terms of k *factors:* i.e., classification variables taking a finite set of values, which we may represent as $1,2, \cdots, m_j$ for the jth factor. In a *complete factorial design* each of the $\prod_{j=1}^{k} m_j$ possible combinations appears just once. For this design there is a simple algorithm based on an orthogonal transformation, that transforms Y into orthogonal components corresponding to the various main effects and interactions which may be defined. The algorithm is a generalization of one given by Yates (1937) for the case that

$$m_1 = m_2 = \cdots = m_k = 2$$

Published algorithms for the basic transformation exist by Claringbold (1969) or Howell (1969) in FORTRAN and by Oliver (1968) or Gower (1969a) in ALGOL60.

Suppose the observed data are Y. Then we can regard the elements of Y as a k-way array with $Y[i_1, \ldots, i_k]$ the response corresponding to level i_j for the jth factor.

The Yates algorithm applies k transformations to the original data. The jth transformation replaces each element in the array by a linear combination of elements differing only in the jth subscript. Specifically, let $Y_0 = Y$ and $s = [s_1, \ldots, s_k]$. Then

$$Y_j[s] = \sum_{i_j=1}^{m_j} M_j[s_j, i_j] Y_{j-1}[s_1, \ldots, i_j, \ldots, s_k],$$

After k steps, Y_k is the output of the algorithm. The m_j by m_j matrices, M_j, are *contrast* matrices. They must be orthogonal and have the property that the first row is constant.

A full analysis of the detailed method is beyond our scope. Criteria of merit for general use of algorithms might include the following. First, the programs should be accurate: the basic algorithm is a series of k orthogonal transformations that should accumulate only a small error due to rounding. Second, the programs themselves should be compact and require little auxiliary storage. Third, the number of arithmetic operations should be kept to a minimum. The latter two conditions can be facilitated by using the *Helmert* matrix for M_j.

One form of this matrix defines

$$M_1[i,j] = \begin{cases} -m_1^{-1/2} & i < j \\ (m_1 - i) m_1^{-1/2} & i = j \\ 0 & i > j \end{cases}$$

for $i = 2, \ldots, m_1$, and similarly for $M_2, \ldots, M_k$. This form requires no extra storage for M_1 (except for $m_1^{-1/2}$) and only about $\frac{1}{2} m_1^2$ additions and $2m_1$ multiplications to multiply M_1 by a vector.

A fourth criterion, not met by any of the referenced algorithms, is that the user be given a simple and general way to unscramble the sums of squares and degrees of freedom. A simple technique is as follows. Define a vector, ψ, of length N such that $\psi[s]$ is the interaction to which $Y_k[s]$ should be assigned. A single pass over Y_k, adding $(Y_k[s])^2$ and 1 to the correct sum-of-squares and degrees-of-freedom term then gives the result.

The definition of ψ comes from representing each of the interactions as an integer. Let ϕ represent any term in the model. Then ϕ is represented as the integer p, where

$$p = \sum_{j=1}^{k} \delta_j 2^{j-1}$$

with $\delta_j = 1$ if ϕ includes the jth factor and $\delta_j = 0$ otherwise. (for example, the grand mean has $p = 0$, the main effects have $p = 1, 2, 4$, etc. and the interaction of the first two factors has $p = 3 = 1 + 2$.) The following ALGOL-style loop computes ψ (the arrays, s and ψ, and the scalar, p, are integers):

```
for j:=1 until k s[j]:=1;

for i:=1 until n
  { for j:=1 until k
    { s[j]:=s[j]+1
      if(s[j]>m[j]) then
        {s[j]:=1; p:=p−2↑(j−1) }
      else if (s[j]=2) then
        p:=p+2↑(j−1)  }
  ψ[i]:=p }
```

Other algorithms for the analysis of variance include the classical method of forming marginal means (Gower, 1969) and the method of sequentially removing effects (Wilkinson, 1970; James and Wilkinson, 1971). These algorithms are not based on orthogonal transformations of the data but may have advantages of simplicity, particularly in special cases or on very small machines.

Further discussion of the mathematical basis of the sequential algorithm is found in James and Wilkinson (1971) and Chambers (1975a). A class of algorithms has been developed also for designs that would be balanced except for the deletion of a few rows from the data matrix. Somewhat similar in treatment are designs that would be balanced except for unequal weighting (unequal replication) of the various factor combinations. An iterative scheme for the former case due originally to Yates (see Preece, 1971) has been adapted in various ways. Rubin (1972) gives a noniterative algorithm and Hemmerle (1974) adapts the iterative method to the unequal replication case.

The nonnumerical facilities of a program for the analysis of variance may contribute significantly to its ease of use: for example, in specifying the model symbolically. Algorithms for this process are not currently available for general use. Fowlkes (1969) and Wilkinson and Rogers (1973) describe some of the important requirements.

m. Summary and Recommendations

The essential algorithmic requirement for linear models is a convenient orthogonal decomposition for regression. Either the Householder or Gram-Schmidt methods are suitable for a general procedure. For problems involving uncertain rank or considerations of errors in the data, the singular-value decomposition is preferred. This is also the basic tool for the classical problems in multivariate analysis discussed in Section **k**. When regression is done by orthogonal decomposition, the complete results may be summarized by the orthogonal coefficients, $Q' \cdot Y$, the upper-triangular matrix, R, and some representation of the orthogonal basis, Q. For reasons of numerical stability, it is not desirable to use the coefficients, B, as a basis for further calculations. Similarly, quantities such as the variance matrix of B or pseudo-inverses of X are not recommended for further calculations.

The recommendation of a standard decomposition is slightly less simple. The Householder method is the simplest accurate method to give the complete description, including a reliable estimate for Q. In addition, it gives a representation for Q_*, the complete orthogonal transformation, and therefore can be used to compute $Q_*' \cdot Y$, leading in turn to accurate residuals. The disadvantage is that current algorithms for this method lack easy-to-use auxiliary routines to compute such quantities.

The Gram-Schmidt method requires reorthogonalization if Q must be accurately orthonormal. The method is not convenient if one requires the complete transformation, but, because Q is returned explicitly, some auxiliary calculations are simplified. The most thorough, complete and flexible set of published algorithms are in Daniel et al. (1976) in ALGOL60, including reorthogonalization and updating. None of the published FORTRAN algorithms is as satisfactory; Clayton (1971) gives the basic algorithm, but a good FORTRAN version of the previous reference would be preferrable.

For those who prefer Householder to Gram-Schmidt, the basic algorithm is Businger and Golub (1965) in ALGOL60. A FORTRAN version, with extensions, is given in Lawson and Hanson (1974). The extensions and much of their discussion involve a rather elaborate technique for transforming on the right by further reflections in the case of apparently deficient rank. This seems of minor relevance in most data analysis applications, but the increased cost is not serious for full-rank data. They give an example illustrating computation of the variances for the coefficients, but do not provide algorithms for the statistical information. Very reliable algorithms based on the Householder procedure are included in or planned for ROSEPACK and LINPACK (see the Appendix).

The use of Givens rotations is a reasonable additional algorithm when one cannot or does not want to handle all the data at one time. It also is the most suitable basic method for sparse problems in which many of the elements of X are exactly zero. Chambers (1971) gives a FORTRAN procedure to compute C and R. Gentleman (1974) gives ALGOL60 procedures, returning as well coefficients and sums of squares. The latter algorithm is generally recommended. The essential procedures are *include* which does one row of the decomposition and *regress* which is a back-substitution for the coefficients. The remaining procedures are either concerned with the special case of deficient rank or involve the rather specialized input/output style for ALGOL60.

The singular-value decomposition algorithms all use essentially the same method, described originally by Golub and Kahan (1966). The recommended algorithms are Golub and Rheinsch (1970) in ALGOL60 and the FORTRAN subroutine SVDRS in Lawson and Hanson (1974).

Published ALGOL60 algorithms for iterative improvement are presented in Björck (1968) and Björck and Golub (1967) for Gram-Schmidt and Householder, respectively. The latter appears in an Argonne Laboratory report (ES257) in a FORTRAN translation. The algorithms are very general, providing for linear constraints and arbitrary rank. As noted in Section **f**, the method has attractive theoretical properties if high accuracy is important, although we have not been able to demonstrate its practical benefits, when higher precision computations are an alternative.

In general, the computations for linear models are well understood and good numerical algorithms are available. The non-numerical aspects of the algorithms are less satisfactory. Most algorithms do not produce all the auxiliary information needed and at the same time do not give the user all the necessary basic summaries and operations to compute these results independently. Statistical packages vary greatly in the amount of information provided, as well as in the control given to the user over the output. Generally, more printed information tends to be given than with the published algorithms. Users concerned with numerical quality should compare the method used with the comments made in

Section **g** concerning accuracy and cost. (Packages should, but sometimes do not, provide an accessible description of method.)

For robust regression, the method of iteratively weighted least squares is convenient and adequate in many examples. It should be emphasized that a full numerical treatment of methods for this problem is still lacking.

The Yates algorithm provides the basic tool for the analysis of variance. See the Appendix for published programs. However, this area still awaits a useful, complete, and reasonably general algorithm.

Problems.

1. (This problem assumes the availability of some program for doing regression.) The following data were used by Longley (1967) and many subsequent writers to examine least-sqaures algorithms:

X						Y
GNP Deflator	GNP	Unem-ployed	Armed Forces	Pop'n	Year	Em-ployed
83.0	234.289	235.6	159.0	107.608	1947	60.323
88.5	259.426	232.5	145.6	108.632	1948	61.122
88.2	258.054	368.2	161.6	109.773	1949	60.171
89.5	284.599	335.1	165.0	110.929	1950	61.187
96.2	328.975	209.9	309.9	112.075	1951	63.221
98.1	346.999	193.2	359.4	113.270	1952	63.639
99.0	365.385	187.0	354.7	115.094	1953	64.989
100.0	363.112	357.8	335.0	116.219	1954	63.761
101.2	397.469	290.4	304.8	117.388	1955	66.019
104.6	419.180	282.2	285.7	118.734	1956	67.857
108.4	442.769	293.6	279.8	120.445	1957	68.169
110.8	444.546	468.1	263.7	121.950	1958	66.513
112.6	482.704	381.3	255.2	123.366	1959	68.655
114.2	502.601	393.1	251.4	125.368	1960	69.564
115.7	518.173	480.6	257.2	127.852	1961	69.331
116.9	554.894	400.7	282.7	130.081	1962	70.551

(a) Regress Y on the X variables, including a vector of 1's for the intercept term. Compute the residuals, Z.

(b) Repeat (a), but remove the means from Y and from the columns of X before the regression, and divide each column by its vector norm (or standard deviation).

(c) Compare the computed residuals. If possible, compute the condition number of the augmented X matrix in (a) and the X matrix in (b). Comment.

(d) Informal opinions from econometricians suggest that the economic variables in the regression generally have uncertainties of about 1%. Assuming population has an uncertainty of about 0.1% and that the year variable enters exactly, adjust the regression for uncertainties in the data as discussed in Section **b**. Hint: Since the year has no error, the regression may be done on residuals of both Y and the remaining columns of X from the year variable.

2. If Z are the residuals from a least-squares regression, show from (3) to (7) of Section **f** that

$$Z' \cdot X = 0$$

i. e., that the residuals are orthogonal to X. Compute $Z' \cdot X$ for the regressions in (a) and (b) and for the adjusted regression in (d) of the previous problem. Compare the accuracy as measured by this test.

3. Weighted Regression:

(a) Derive the equations for weighted regression with intercept in **h**.

(b) Show by example that the condition number for X_W can be as large as

$$\kappa(X_W) = \kappa(W) \times \kappa(X)$$

(Note that $\kappa(W) = \max w_i / \min_{w_i > 0} w_i$.)

(c) How would you interpret and handle an example of robust regression where the condition of the original, unweighted X was substantially better than that of the final X_W using the robust weights?

CHAPTER SIX

Nonlinear Models

a. Introduction

This chapter considers computational techniques for fitting and to some extent for assessing nonlinear models. A nonlinear model, for us, is a model for data that cannot be fitted by the linear techniques of Chapter 5. This includes many models that are linear in form but do not use least squares. Computationally, we are primarily interested in general, iterative methods for problems such as optimization, nonlinear equations and nonlinear least squares.

Nearly every aspect of nonlinear fitting requires more care than in the case of linear models. The numerical properties of the various methods are more difficult to assess in theory, and the theory may be a poorer prediction of behavior in practice. Published or generally available algorithms are fewer in number and sometimes less satisfactory in care and completeness. At the same time, the user must be careful when writing his own algorithm, since details are of significant importance.

These negative remarks are not intended to discourage the use of nonlinear models. On the contrary, the careful use of general model-fitting techniques can greatly increase the scope of data analysis. However, the belief that it is only necessary to set the computer iterating to obtain a good solution to any model on any data is bound to lead the user into painful experiences.

In this chapter we present a few of the most important numerical methods. Emphasis is given to those methods that provide the best combination of numerical performance and useful statistical information. These are primarily the methods based upon quadratic approximation of an objective function, including the special case of nonlinear least squares. Other methods described may be preferred in special cases (e.g., models with very many parameters or an unsmooth objective function), or may appeal to users for nonnumerical reasons (e.g., they are easy to program).

To begin, we define a general model-fitting problem. Suppose that N observations, say $Y = (y_1, \cdots, y_N)$, have been obtained. We are concerned with applying to these data a class of models, generally formulated mathematically, indexed by an m-dimensional parameter, θ. The model-fitting problem is then to select a value or a set of values for θ for which the model is a good representation of the data, in some chosen sense.

The statistical aspects of the model often include some probabilistic mechanism. In any case, the data analysis must provide a specific link between data and model before numerical fitting may proceed. Three classes of models in this respect are commonly used:

(i) **Predictive models.** For each of the N observations, the model suggests a predicted value, depending on the parameter, and possibly on other data values. The most general means of expressing this is in terms of N functions $f_1(\theta), \cdots, f_N(\theta)$ such that $f_i(\theta)$ predicts y_i.

(ii) **Probability models.** The model assigns, given values for Y and θ, a probability element, say in terms of a probability density function $p(Y;\theta)$.

(iii) **Transformation models.** In this case the data as a whole are replaced by transformed values $U = (u_1, \cdots, u_n)$, where the transformation depends on a parameter, θ, and the objective is to find U with certain desired properties.

The model definition is completed by specifying a criterion for an acceptable set of parameters. In most cases this criterion attempts in principle to define a unique "best" set, $\hat{\theta}$; for example, by choosing $\hat{\theta}$ to minimize some *objective function,* $F(\theta)$.

Examples of objective functions for classes (i), (ii) and (iii) are, respectively, the sum of squared residuals, minus the log likelihood function and a badness-of-fit function in multidimensional scaling (Kruskal and Carroll, 1969).

Other criteria for best estimates may be framed directly as a set of (nonlinear) equations that may be stated in the form,

$$g(\hat{\theta}) = 0. \tag{1}$$

This formulation is closely related numerically to optimization when the elements of g are taken to be the partial derivatives of F with respect to θ. In this chapter, most of the discussion uses the optimization formulation. Section **h** discusses some of the distinctions. A special form of nonlinear equation is the *fixed point method* in which one rearranges (1) in the form

$$\hat{\theta} = \Gamma(\hat{\theta}).$$

This has been popular in many problems because it suggests the obvious algorithm of putting an initial estimate, θ_1, into the right side, obtaining a new estimate, θ_2, from Γ and proceeding iteratively until θ_j converges.

A class of estimation methods has also been proposed which truncates an iterative procedure after one or more (but a fixed number) of steps. See Bickel (1975) for an example in robust linear fitting. The argument in favor of these methods is presumably one of computational efficiency. Mathematically, however, they suffer from the difficulty of saying much about the properties of the estimates and the dependence of the results on starting values and on details of the computation.

b. Optimization; Quadratic Methods; Newton-Raphson

The most active work on methods of numerical optimization has been based on the use of an approximating form for the objective function, and by far the majority of this work uses a quadratic function for approximation. The minimum of the approximation suggests an estimate of the minimum for the actual objective function. If the estimate is not adequately close, a new approximation is computed (using the function value and possibly the derivatives at the estimate) and the process repeated.

Quadratic approximations have several advantages for optimization. First, the minimum of the approximation can be computed by relatively simple calculations. Second, one suitable method of quadratic approximation, the Taylor series, has been known for centuries and others have been developed more recently. Third, if the approximation is sufficiently accurate near the current parameter estimate, the method should exhibit accelerated convergence in the limit (see Section **e**), so that high numerical accuracy is possible. Fourth, the quadratic approximation provides asymptotic estimates of variance for many common models.

In general, a quadratic approximation to $F(\theta)$ about the point θ_0 may be written:

$$F_0 + g_0 \cdot (\theta - \theta_0) + \tfrac{1}{2}(\theta - \theta_0) \cdot G_0 \cdot (\theta - \theta_0) \qquad (2)$$

where F_0, g_0, and G_0 are a scalar, a vector, and a symmetric matrix.

The *gradient* or vector of first derivatives at θ is

$$g_0 + G_0 \cdot (\theta - \theta_0)$$

and the *Hessian* or matrix of second derivatives is G_0. The minimum of the approximation, if any exists, must have a zero gradient, so that $\theta_1 = \theta_0 + \delta_0$ is a minimum only if δ_0 satisfies the linear equations

$$G_0 \cdot \delta_0 = -g_0. \tag{3}$$

If G_0 is positive definite (i.e., if $\delta \cdot G_0 \cdot \delta > 0$ for all $\delta \neq 0$), then the solution to (3) is unique, and represents a minimum of the quadratic.

Two methods for constructing quadratic approximations, one old and one much more recent, dominate computational methods. The *Newton-Raphson method,* in essence, defines g_0 and G_0 in (2) to be the vector of first derivatives of F and the matrix of second derivatives evaluated at θ_0. From (3) a new estimate, $\theta_1 = \theta_0 + \delta_0$, is defined, new derivatives g_1 and G_1 computed, and the process repeated.

For computational purposes, the method needs to be refined. Both in theory and in practice, the iteration may easily fail to converge. The second-derivative matrix need not be positive definite, so that even the approximation may not have a local minimum. Even if G_0 is positive definite, it does not follow that θ_1 is a better estimate of the minimum, in the sense that $F(\theta_1) < F(\theta_0)$.

A variety of remedies have been suggested to *stabilize* the method. A general principle underlying many of them, as well as other methods in optimization, is the following. Let δ_0 be a correction in (3) pointing downhill; i.e.,

$$\delta_0 \cdot g_0 < 0, \tag{4}$$

where g_0 is the gradient of F. If F is sufficiently regular (specifically, continuously twice differentiable), and θ_0 is not already a minimum, then for $||\delta_0||^2/|\delta_0 \cdot g_0|$ sufficiently small,

$$F(\theta_0 + \delta_0) < F(\theta_0). \tag{5}$$

The mathematical derivation uses the first-order Taylor series for $F(\theta_0 + \delta_0)$, with a remainder which is of the order of $||\delta_0||^2$. Then if $||\delta_0|| < -\epsilon \delta_0 \cdot g_0$ with $\epsilon > 0$,

$$F(\theta_0 + \delta_0) - F(\theta_0) < \delta_0 \cdot g_0 (1 - \epsilon^2 C) < 0$$

for ϵ small enough. Algorithms requiring (5) to hold for each step are called *descent* methods.

Stabilized Newton-Raphson algorithms, most quasi-Newton methods, and various stabilized nonlinear least-squares algorithms use special forms for this result. For example, let H_0 be a fixed, positive-definite matrix and define δ_0 by

$$\delta_0 = -\alpha H_0 \cdot g_0 \tag{6}$$

for positive scalar α. Then for α small enough $F(\theta_1) < F(\theta_0)$. A second special case is to define δ_0 by

$$(G_0 + \alpha^{-1} I) \cdot \delta_0 = -g_0 \tag{7}$$

where G_0 is fixed (not necessarily the Hessian). Again, for α sufficiently small, $F(\theta_1) < F(\theta_0)$. Both these examples are used in practice. Merely ensuring an improvement on each step does not, in itself, guarantee a good algorithm; much of the detailed algorithmic design goes to ensure that the step is a *significant* improvement.

A class of stabilized Newton-Raphson methods uses corrections similar to (7). Suppose λ_m is the smallest eigenvalue of G_0. If $\lambda_m < 0$, G_0 is not positive definite; but if we take $\alpha < -1/\lambda_m$, the matrix in (7) is positive definite. The general idea is to choose such an α, evaluate δ_0, and then compute $F(\theta_1)$. If $F(\theta_1) < F(\theta_0)$ we go on to the next step. Otherwise one may either repeat (7) with a smaller α or search for a point, $\theta_0 + c\delta_0$, for $0 < c \leqslant 1$ at which F is reduced (e.g., by cubic interpolation, as in the quasi-Newton methods). The general result guarantees that both methods work. Goldfeld, Quandt and Trotter (1966) describe an algorithm of the first type.

c. Quasi-Newton Methods

The second widely used class of quadratic approximations differs from Newton-Raphson in two important ways: second derivatives of F are not computed, and the quadratic approximation is built up gradually from one step to the next. The basic idea was proposed by Davidon (1959), developed by Fletcher and Powell (1963), and elaborated upon and varied by many workers in the following ten years or so.

Davidon observed that it was possible to determine the minimum of a quadratic function like (2), using only function values and first derivatives. This can be done by defining a sequence, $H_0, H_1, \cdots$, of matrices, setting $\theta_{i+1} = \theta_i + \delta_i$, with

$$\delta_i = -\alpha_i H_i \cdot g_i$$

and g_i the gradient at θ_i. Two requirements are placed on δ_i. First, α_i is chosen to give the minimum of the quadratic function of α, $F(\theta_i - \alpha H_i \cdot g_i)$. Second, H_i is defined so that

$$\delta_{i-1} = H_i \cdot (g_i - g_{i-1}). \tag{8}$$

These two conditions are sufficient to ensure that θ_m is the minimum, assuming F is quadratic and positive definite. The original method defined H_i by a correction to H_{i-1} of rank two. Let $\gamma = g_i - g_{i-1}$, $\eta = H_{i-1} \cdot \gamma$, and write $x \otimes y$ for the matrix with $x_r y_s$ as the element in the rth row and s^{th} column. Then

$$H_i - H_{i-1} = \frac{\delta_{i-1} \otimes \delta_{i-1}}{\delta_{i-1} \cdot \gamma} - \frac{\eta \otimes \eta}{\eta \cdot \gamma} \tag{9}$$

Since the Newton-Raphson method finds the minimum of a quadratic in one step, intuition suggests that m steps of a quasi-Newton method should be similar to one Newton-Raphson step. There is some practical and mathematical evidence in favor of this (see Section **e**). There is no definitive rule preferring one method or the other. Some of the quasi-Newton method's appeal is that less user programming may be necessary if the user is responsible for differentiating the objective function. See also Section **j**.

The original quasi-Newton method has appeared in several published ALGOL60 algorithms. Wells (1965) has serious errors, but corrections by Fletcher (1966) and Hamilton and Boothroyd (1969) were published. Lill (1970) implements the method using difference approximations to derivatives.

During the decade or so following the publication of the paper by Fletcher and Powell, a flood of material on quasi-Newton methods appeared. Aside from papers looking into the mathematical properties of the methods (which we treat briefly in Section **e**), there were many suggestions for alternative procedures. These focused on two aspects: the search along the direction, $-H_i \cdot g_i$, and the procedure for updating H_i.

Davidon's original argument required that δ_i provide (at least) a quadratic approximation to the minimum of the one-dimensional function, $F(\theta_i - \alpha H_i \cdot g_i)$. To do so required the evaluation of F and its derivative for two values of α, at least. This contrasts unattractively with the Newton-Raphson method, where the *direct* step, $\alpha = 1$, gives the correct value for a quadratic function. Although one choice of correction to H_i (the "rank one" formula) does provide convergence on a quadratic without minimizing for α, this method's popularity was relatively brief. In practice, it sometimes produces H_i that are not positive definite, and in general may behave in an unstable way, unless care is taken in the choice of step length. However, many updating formulas can be used if we relax the requirement that the method converge exactly in m steps on a quadratic. The step length is usually required to give a *significant* improvement in F (but not always a minimum) and to satisfy some conditions for good updates of the quadratic approximation. An attractive method due to Davidon (1975) maintains quadratic termination *without* line searches. The technique is to use a projection of the search direction to compute updates for H_i, such that (8) is satisfied.

The updating formulas for H_i were also generalized. General corrections to H_i of rank two were derived as parametric families that included the original (9) as a special case. One such parametric family was defined by Broyden (1967) and later by Fletcher (1970) and Shanno (1970). Many of the successful quasi-Newton methods used members of this family. A theoretical result of Dixon (1972) showed that all these methods would be exactly alike on any differentiable function if α_i were chosen to give the exact minimum. In practical terms, this suggests that the methods will behave similarly if minima are approximated along the line and that, in any case, the choice of α_i will significantly affect performance. Dixon (1972a) compared a number of such methods on a battery of twelve test functions. The major conclusion was that the modified methods had a general advantage over the original method of about 25 to 50%.

Other important modifications have been made. By analogy with the computation of linear least squares from the decomposition of the data rather than from the cross-products, it has been suggested that the Hessian approximation could be computed in terms of a decomposition; for example,

$$H_i = R_i' \cdot R_i$$

with R_i an upper-triangular matrix. Working with the factor rather than the Hessian should improve the numerical conditioning of the calculation, reducing the effect of rounding in updating. Gill and Murray (1972) and Davidon (1975) propose methods of this form. A related problem is the scaling of the parameters. While the model is unaffected by multiplying the parameters by arbitrary constants and adjusting derivatives accordingly, the effect on a quasi-Newton method may be considerable, particularly in the initial stages. Attempts have been made to include scaling in the algorithms, e. g., Davidon (1975), but the problem remains difficult and important.

The requirements for a user of a quasi-Newton algorithm are similar to those for a Newton-Raphson algorithm: the writing of a subprogram to evaluate the particular objective function and its first derivatives (unless these are approximated as in the algorithm of Lill) and, usually, the specification of a desired accuracy for convergence. Particular algorithms may make additional requests; for example, in Fletcher (1970) an estimate of the function value at the minimum is needed. In Davidon (1975) an initial estimate of H (actually a factorized version of H) is strongly recommended.

Accumulated empirical and theoretical experience suggest that one of the modified quasi-Newton methods should be a basic routine for model fitting. Of published, refereed algorithms, Shanno and Phua (1976) would appear to be the best recommendation. However, this method does not include the more recent techniques, such as factoring the Hessian. The ideas in Davidon (1975) seem promising, but at the time of writing, no published algorithm based on these exists. See the Appendix for further algorithms.

d. Other Optimization Methods

Many techniques have been proposed, but the most influential, in addition to Newton and quasi-Newton, fall into two groups:

(i) search methods not requiring derivatives;

(ii) methods whose storage requirements are linear in the number of parameters.

Methods of the first group generally involve looking along search directions for improvement in $F(\theta)$ and maintaining, by some mechanism, a basis of m directions or vectors to guide the search. An algorithm developed originally by Powell (1964) and later extended by Brent (1973) involves an implicit quadratic approximation. Minima are approximated along the search directions (by quadratic interpolation), and the method has finite convergence for quadratic functions. Also, in Brent's form, an approximation to the inverse Hessian is produced. This method is therefore in direct competition with quasi-Newton methods using difference approximations, and Brent cites empirical evidence that it is more efficient than the original Davidon method in this form. An algorithm in a variant of ALGOL appears in Brent (1973).

The older search methods did not use quadratic approximations. This makes the methods less attractive when high accuracy is required. On the other hand, there is some evidence that they may be useful for functions that are "rough"; for example, functions observed with error (Springer, 1971 and later unpublished work). The method of Nelder and Mead (1965) is relatively widely used, with a FORTRAN algorithm by O'Neill (1971) and corrections noted in Chambers and Ertel (1974).

Methods of class (ii) may be suitable when the number of parameters is too great to permit storing an m by m matrix, but when there is presumably some reason to believe that an approximation to a minimum can be computed at an acceptable cost. The oldest and simplest such method is that of steepest descent. From the current estimate θ_i one proceeds along the negative gradient to

$$\theta_{i+1} = \theta_i - \alpha g_i$$

choosing α, for example, to obtain an approximation to the minimum along $-g_i$. This method is still used in some fields (for example, multidimensional scaling), but it does not have accelerated convergence (see Section **e**) and can be extremely slow. The *conjugate gradient* methods introduced by Fletcher and Reeves (1964) are more like quasi-Newton methods. Minimization along m successive directions produces the minimum of a quadratic. This method does have accelerated convergence so that high accuracy is feasible. The methods choose successive search directions, h_i, which are conjugate with respect to F if F is quadratic; specifically,

$$h_{i+1} \cdot G \cdot h_i = 0.$$

No approximation of the quadratic is formed, however; instead, h_{i+1} is computed as a linear combination of the previous h_i and the current gradient. Given θ_i and h_i, one forms

$$\theta_{i+1} = \theta_i + \alpha\, h_i$$

where α is chosen to minimize along the search direction. (For example, one of the search procedures in the algorithms for the Davidon-Fletcher-Powell method could be used.) Then,

$$h_{i+1} = -g_{i+1} + \gamma\, h_i$$

Choices for γ include

$$(g_{i+1} \cdot g_{i+1})/(g_i \cdot g_i)$$

(Fletcher and Reeves, 1964) or

$$((g_i - g_{i+1}) \cdot g_{i+1}) / (g_i \cdot g_i)$$

(Polak and Ribiere, 1969). Initially, $h_0 = -g_0$, and it is essential to *restart* the algorithm by setting $h_i = -g_i$ after every k steps, for k somewhat larger than m. See the comments on Figure 1 in Section e.

The conjugate gradient methods are preferred as general methods for problems too large to store the quadratic approximation.

e. Mathematical Properties

During the early 1970s a number of mathematical results were proven concerning subsets of the class of quasi-Newton methods and some of the conjugate gradient methods. Three main topics are covered: finite termination for a quadratic, convergence on general functions, and rate of convergence. We state here the main results, with at most a sketch of the proofs. (For several of the results, the original proofs are quite difficult.)

The following definitions will be useful. Given a point θ_0 we call the *level set* $L(\theta_0)$ the set of points with at least as small value for F:

$$L(\theta_0) = \{\theta \mid F(\theta) \leqslant F(\theta_0)\} \tag{10}$$

If H is an m by m matrix, we say that H is *bounded-positive-definite* if there exist positive constants c_0 and c_1 such that

$$c_0 \leqslant \lambda_j(H) \leqslant c_1$$

where $\lambda_j(H)$ are the eigenvalues of H. The function $F(\theta)$ is said to be *bounded-convex* on a set if its second-derivative matrix is continuous and bounded-positive-definite on the set. A step in any optimization method is called *perfect* if the step length minimizes exactly the function

$$h(t) = F(\theta + t \times s).$$

The notation, $F \in C^{(r)}$, means that F and its first r derivatives are continuous on a set to be defined. The parameters, θ_0, represent starting values for the computation.

	Newton–Raphson	Quasi–Newton (DFP)	Conjugate Gradient
Termination on Quadratic	In 1 direct step	After m perfect steps if G positive–definite	Same as quasi–newton
Convergence	$F \epsilon C^{(3)}$ and boundedly convex; θ_0 in suitably small neighborhood of $\hat{\theta}$ for direct steps	$F \epsilon C^{(2)}$; boundedly convex on $L(\theta_0)$	$F \epsilon C^{(3)}$, boundedly convex on $L(\theta_0)$
Rate of Convergence	As above; second order	As above and a Lipschitz condition on $G(\theta)$; faster than linear	As above; second order for m steps ahead
References	Polak (1971, p. 254)	Fletcher and Powell(1963); Powell(1971; 1972a)	Polak and Ribiere (1969); Cohen (1972)

Figure 1: Convergence of Optimization Methods

Figure 1 presents a set of convergence results for three important classes of methods: Newton-Raphson, quasi-Newton and conjugate gradient. By Newton-Raphson, we mean the direct prediction method, unmodified. By quasi-Newton we mean the original method with perfect steps. Dixon (1972) shows that a family of quasi-Newton methods are identical, ignoring rounding, so that the results apply more widely, given perfect steps. By conjugate gradient we mean the Fletcher-Reeves or Polak-Ribiere methods, as discussed in Section **d**. Note that the restart mentioned previously is essential for this method. Powell (1976) shows that the convergence is generally not superlinear otherwise. The conditions described in Figure 1 are chosen to be simple and relatively easy to verify. They are not always the broadest conditions and some significant alternative results exist. The general intuition behind the results is somewhat as follows. The three classes of methods will converge in a region where $F(\theta)$ has a single smooth valley, with a unique minimum. If k steps are required to minimize an exact quadratic, then the limiting rate of convergence should be such that

$$||\theta_{i+k} - \hat{\theta}|| = O(||\theta_i - \hat{\theta}||^2).$$

While the theory so far does not quite substantiate intuition on all counts, the results are in general agreement.

The original quasi-Newton method had the property that it converged exactly (ignoring rounding error) for a quadratic function.

The key to this result is the assumption that the step length is perfect in the sense defined, at least for a quadratic. In this case, many of the more recent updating formulas would have finite termination, as would the conjugate direction method and the search method of Powell (1964). The updating correction of rank one has finite termination without perfect steps and Powell (1972) and Davidon (1975) show termination for some other methods with relaxed step requirements, but many quasi-Newton algorithms, such as Fletcher (1970), prefer to abandon finite termination.

The guarantees of convergence illustrate the importance of the step-size strategy. A purely predictive strategy of taking $\alpha = 1$ generally reduces us to local convergence only; e.g., θ_0 must be within a distance, δ, of $\hat{\theta}$ for the Newton-Raphson method to be guaranteed, where δ depends on F and cannot be easily determined. In contrast, if we assume perfect steps, a general convergence result is obtained as in Figure 1, for the Newton-Raphson method as well: if $F \epsilon C^{(2)}$ and F is bounded-convex on $L(\theta_0)$, then the method converges. See the discussion in Polak (1971) for more details.

The same distinction applies to the quasi-Newton methods. The results of Powell (1971; 1972a) combined with Dixon (1972) prove convergence for a broad class of quasi-Newton methods under the conditions shown. However, we have noted that considerable empirical evidence suggests that perfect steps are inefficient in practice, so that one would like theoretical support for a less stringent algorithm. Broyden, Dennis and More (1973) showed such a proof for the direct prediction strategy, but again the result was local and also depended on a condition bounding the error in G_{i+1} as an estimate of $G(\hat{\theta})$. The function, F, is required to be $C^{(2)}$ and to satisfy a Lipschitz condition on G: namely,

$$||G(\hat{\theta}) - G(\theta)|| \leqslant K||\theta - \hat{\theta}||^p \qquad (11)$$

for some $p > 0$.

The conditions were shown to be satisfied by the symmetric rank-one method and by a family of methods which includes the empirically successful algorithms of Fletcher (1970) and Powell (1970a; 1970b). Broyden et al. (1973) also showed that superlinear convergence was observed under these assumptions; namely

$$||\theta_{i+1} - \hat{\theta}||/||\theta_i - \hat{\theta}|| \rightarrow 0$$

as $\theta_{i+1} \rightarrow \hat{\theta}$. The Lipschitz condition required by Powell (1972a) was similar to (11) but with $p = 1$.

f. Distribution of Estimates

An important class of nonlinear models contain some probabilistic component, in the sense that the model assigns to the data, Y, some probability distribution. The optimization problem then is generally a parameter estimation technique in the statistical sense. The parameter estimate, $\hat{\theta}$, is defined as a function of the data, Y, indirectly, by the requirement that

$$F(Y, \hat{\theta}) = \min_{\theta} F(Y, \theta)$$

Therefore, the assumption that Y has some distribution implies that $\hat{\theta}(Y)$ has also a derived distribution. The examination of this distribution and statistical inferences about it are important tools in interpreting the model. In most nonlinear models, the exact distribution of $\hat{\theta}$ is intractable, but several approximate techniques are frequently useful. This section presents computational methods for this purpose. The general statement is as follows. Our model asserts that Y has a probability distribution, say $P(Y, \psi)$, where ψ is some vector of parameters indexing the distributions. Then $\hat{\theta}(Y)$ has a derived distribution, also depending upon ψ. The estimated parameters, θ, and objective function, $F(Y, \theta)$, are nearly always related to the distribution parameters ψ and distribution $P(Y, \psi)$; in fact, the presence of parameters in $P(Y, \psi)$ is usually the reason for doing the estimation. Therefore, θ is either equivalent to or a function of ψ. To understand the use of simulation, however, it is important to keep the distinction in mind, and we will continue to use different symbols for the two parameters.

The *simulation* or *Monte-Carlo* method allows the generation of a sample of M pseudorandom values, $Y^{(1)}, \cdots, Y^{(M)}$, from the distribution, $P(Y, \psi)$, for given ψ. Applying to each of these the minimization of $F(Y^{(j)}, \theta)$ generates a sample, $\hat{\theta}^{(1)}, \ldots, \hat{\theta}^{(M)}$, from the distribution of $\hat{\theta}$. These M values in turn allow the estimation of any property of the distribution, such as mean, variance, or quantiles. By increasing the value of M, we can generally increase the accuracy with which these properties are estimated, at the cost of more computation.

The steps to be taken are more precisely described as follows:

I. Set $\psi = \psi_{gen}$

II. For $j = 1, \cdots, M$

a. Generate a pseudorandom value $Y^{(j)}$ from $P(Y, \psi_{gen})$

b. Compute $\hat{\theta}^{(j)}$ to minimize $F(Y^{(j)}, \theta)$

III. Compute the desired summaries from $\hat{\theta}^{(1)}, \cdots, \hat{\theta}^{(M)}$.

While simulation can only *estimate* properties of the distribution of $\hat{\theta}$, it is in an important sense the only general method offering arbitrarily high accuracy in these estimates. Given certain mild assumptions of regularity in the distribution $P(Y, \psi)$, any specific property can be estimated arbitrarily well by taking M sufficiently large, at least if defects in the pseudorandom generators are ignored. The other estimates to be discussed do *not* have this property, and so must be considered less satisfactory. Generally they are asymptotic in N, the number of observations in Y, in that the error of estimation goes to zero as N becomes large. The methods offer some compensation, however, in generally requiring less computation. The results obtained use only the *observed* data, and from now on we assume $Y = Y_{obs}$.

We also write θ_T for the value of θ obtained by setting ψ to some fixed value, ψ_T.

The classical estimate of variance in maximum-likelihood estimation is due to Fisher (1922). Suppose $F(Y, \theta)$ is the negative log-likelihood, with $g(y, \theta)$ and $G(Y, \theta)$ the first and second derivatives. Fisher's result is that the inverse of $G(Y, \hat{\theta})$; i.e., the matrix we have denoted by $H(\hat{\theta})$, provides an estimate for the variance of $\hat{\theta}$. The precise conditions are as follows. Suppose $Y = [y_1, ..., y_n]$ is a sample of N from the distribution having density $p(y, \theta_T)$ (we now assume that θ determines, at least, the distribution of $\hat{\theta}$).

We say that the distribution of y is *regular* if we can differentiate twice across the integral sign in the identity

$$\int p(y, \theta)\, dy = 1; \tag{12}$$

To express this result we define

$$q(y, \theta) = \partial \log(p(y, \theta))/\partial\theta$$

where $p(y, \theta) > 0$, and similarly

$$Q(y, \theta) = \partial q(y, \theta)/\partial\theta$$

Then p is regular at $\theta = \theta_T$ if and only if

$$E(q(y, \theta_T)) = 0$$

$$E(Q(y, \theta_T)) + E(q(y, \theta_T) \otimes q(y, \theta_T)) = 0 \tag{13}$$

where E( ) denotes expected value when $\theta = \theta_T$. It is assumed all expectations exist and are finite in (13). The verification comes directly from twice differentiating (12). Fisher's result is then precisely stated as:

> *Theorem.* If p is regular for θ in a neighborhood of θ_T, if Y is a sample from $p(y, \theta_T)$, and if the maximum likelihood estimate $\hat{\theta}$ is unique for N sufficiently large, then the statistic, $\hat{\theta}$, has asymptotically a multivariate normal distribution with mean, θ_T, and variance matrix, $H(\theta_T)$. In particular, $H(\hat{\theta})$ estimates the variance of $\hat{\theta}$.

The result is only valid asymptotically, so that the exact numerical values should be used with caution. As an indication of magnitude, the results can be quite useful. The great advantage is that many of the optimization procedures discussed produce an estimate of $H(\hat{\theta})$ automatically, even when no explicit calculation of second derivatives takes place.

Other models besides maximum likelihood satisfy the conditions for asymptotic normality. Under similar regularity conditions, parameters estimated by nonlinear least squares can use the same asymptotic variance estimates (Jennrich, 1969). Ertel (1975) and Crowder (1976) have generalized the assumptions beyond the case of independent observations. The former applies the theory to some common classes of robust estimates (see Section **5.j**).

The results of this section suggest an advantage of Newton and quasi-Newton methods, separate from considerations of numerical efficiency.

It should be kept in mind that the conditions of the theorem are *not* always satisfied. If the model is such that θ defines the region in which $p(y, \theta) > 0$ (for example, a uniform distribution on $[\theta_1, \theta_2]$), we cannot exchange integral and derivative. The expectations in (13) may not exist; for example, in a Cauchy family,

$$p(y, \theta) = 1/(1 + (\theta_1 + \theta_2 y)^2).$$

In many models, the y_i are not precisely identically distributed. While the asymptotic estimates may be helpful even when the assumptions are known to be violated, they should always be used with caution, not as a statistical black box.

g. Nonlinear Least-Squares Estimation

The class of special nonlinear models most commonly treated are least-squares predictive models:

$$F(Y,\theta) = \sum_{i=1}^{N} (y_i - f_i(\theta))^2 \tag{13}$$

In this case f_i is the (nonlinear) regression function to predict y_i. Special-purpose routines for this problem begin with a local linear approximation to the model; say,

$$f_i(\theta + \delta) \approx h_i + a_i \cdot \delta \tag{14}$$

where the scalar, h_i, and vector, a_i, depend on θ (and possibly on other information as well). Substituting (14) into (13), one chooses δ to be the solution of the linear least-squares problem:

$$\min_{\delta} ||z - A \cdot \delta||^2 \tag{15}$$

where $z_i = y_i - h_i$ and a_i is the ith row of A. The classical Gauss-Newton method is precisely this, with (14) defined by a Taylor series:

$$h_i = f_i(\theta); \quad a_i = \partial f_i / \partial \theta.$$

Given a good linear least-squares algorithm, this method can be implemented easily.

The linearized approximation in (14) can be compared to the quadratic approximation in the Newton-Raphson method. The first and second derivatives of F in (13) are:

$$g(Y,\theta) = -2\sum_i (y_i - f_i) \cdot a_i$$

$$G(Y,\theta) = 2A' \cdot A - 2\sum_i (y_i - f_i) \cdot \partial a_i / \partial \theta$$

where A is the N by m matrix whose i^{th} row is a_i. The expression for $G(Y,\theta)$ contains two terms: the first is what G would have been had the model actually been linear; the second corrects this by a sum of matrices weighted by the residuals, $y_i - f_i(\theta)$. The Gauss-Newton correction, by virtue of being a linear least-squares solution, satisfies

$$\begin{aligned} 2(A' \cdot A)\delta &= 2A' \cdot (y_i - f_i) \\ &= -g \end{aligned} \tag{17}$$

while the quadratic approximation gives a step δ_q with

$$G \cdot \delta_q = -g.$$

Thus the two corrections satisfy equations with the same right side and left sides differing by the second term in the expression for G. This term will vanish if the model is linear or if the residuals at θ all vanish. The implication is that, when the second term is relatively small, Gauss-Newton behaves like Newton-Raphson.

By assuming that f_i has continuous higher-order derivatives and finding limiting expressions for $g(\theta + \delta)$ one may make this statement more precise and show that

$$||\theta_{i+1} - \hat{\theta}|| = O(||\theta_i - \hat{\theta}||^2) + O(||\theta_i - \hat{\theta}||\ ||\hat{z}||)$$

with $\hat{z}_i = y_i - f_i(\hat{\theta})$ the residuals at $\hat{\theta}$. This should be compared to the table in Section **e** and the difference noted: The Gauss-Newton method will *not* have quadratic convergence beyond some point at which $||\theta_i - \hat{\theta}||$ is no longer large compared to $||\hat{z}||$. There is a similar flavor to the result, quoted in Section **5.g**, relating the accuracy of *linear* least squares to the condition of X.

Stabilizing modifications have been proposed for the Gauss-Newton method, as for the Newton-Raphson. Notice from (17) that if A is nonsingular, δ is the product of a positive-definite matrix times $-g$ and therefore is a descent direction. Hence F must be improved by a step, $\alpha\delta$, for some scalar $\alpha > 0$. Suggested methods for determining α have existed for many years (e.g., Hartley, 1961), but a very simple one would be to use cubic interpolation along δ, as in the quasi-Newton algorithms, with the necessary gradients defined by (16).

A more popular approach has been to redefine (17) as the criterion for δ. The most common method has modified (17) to

$$(A' \cdot A + \Lambda)\delta = A' \cdot z, \qquad (18)$$

where Λ is a positive-definite matrix, often λI for some positive scalar λ.

Algorithms of this form select a correction Λ_i at step i such that the solution, δ_i, to (18) at θ_i gives an improved sum of squares. This can always be done, for example, if λ is made very large, since then δ_i will be nearly a small step in the steepest descent direction. On the other hand, the methods attempt to make $\Lambda \rightarrow 0$ as the algorithm nears a solution. If this can be done, the limiting behavior should be like the unmodified Gauss-Newton method.

Methods of the form of (18) are similar to some attempts to stabilize linear regression (see, for example, Marquardt, 1970). They are all most easily regarded as replacing X or A by a matrix that is better conditioned numerically. Many variants are possible in either case, augmented in the nonlinear case by varying strategy in altering Λ. Good, generally available algorithms have been rare. An implementation of Marquardt (1963) circulated in the 1960s through the SHARE organization. Roger Fletcher's FORTRAN algorithm, with a modified strategy on changing Λ, appeared in a U.K. Atomic Energy Research Establishment technical report, R7125.

The modified Gauss-Newton methods can significantly improve the initial behavior of the method. They cannot improve the limiting behavior nor avoid the difficulties that arise because the residuals from the model are large. A promising approach for the latter problem has been to consider the difference between the Hessian, G, and its linearized approximation, $A' \cdot A$, in (16). An approximation to G is developed, similar to the quasi-Newton approximations, but approximating only the *difference*. An algorithm has been developed along these lines by J. E. Dennis, D. Gay, and R. E. Welsch at National Bureau of Economic Research (see ROSEPACK in the Appendix).

A frequently encountered special classs of nonlinear least-squares problems are *separable* or *partially linear:*

$$\theta = (\theta_{(1)}, \theta_{(2)})$$

such that $f_i(\theta)$ is a linear function of $\theta_{(2)}$ given $\theta_{(1)}$. Techniques can then be developed which fit $f_i(\theta_{(1)}, \hat{\theta}_{(2)}(\theta_{(1)}))$ taking the linear solution $\hat{\theta}_{(2)}(\theta_{(1)})$ as given. This has the advantage of reducing the number of parameters in the nonlinear problem. Some commonly used and numerically difficult problems fall in this class; e. g.,

$$f_i(\theta) = \sum_j \theta_{k+j} e^{\theta_j}$$

For such examples, a marked improvement in numerical performance has been observed when the linearities are exploited. In order to exploit the linear terms with an algorithm requiring derivatives, one must compute the derivatives of the linear projection operator, $\partial \hat{\theta}_{(2)} / \partial \theta_{(1)}$. See Golub and Pereyra (1973) and Kaufman (1975).

h. Nonlinear Equations; Fixed Point Methods

The treatment of optimization as the basic operation in this chapter is somewhat arbitrary, chosen partly for convenience in leading to discussions of constrained optimization and nonlinear least-squares. Otherwise one could take as the fundamental process the solution of a system of nonlinear equations.

$$g(\theta) = 0 \tag{19}$$

where $g = (g_1(\theta), \cdots, g_m(\theta))$ is a vector of m real-valued functions of the real m-vector, θ.

The solution of (19) is related to the general minimization problem in several ways. If $\hat{\theta}$ is a local minimum of a differentiable function $F(\theta)$, then $\hat{\theta}$ satisfies (19) with $g(\theta)$ the gradient of $F(\theta)$. Conversely, one may, in principle, replace (19) with the minimization of any function, $L(g(\theta))$, where $L(y)$ is a real-valued function of m variables with a unique minimum at $y = 0$. For example, $L(y) = ||y||^2$ converts (19) into the minimization of

$$||g(\theta)||^2 = \sum_i (g_i(\theta))^2.$$

This has some advantage in producing a well-defined answer even if (19) has no exact solution.

Many of the numerical techniques for nonlinear equations are close parallels to minimization techniques. In particular, the entire class of quadratic approximation methods have the obvious analogues in (19) of correcting the current θ by a value, $\theta + \delta$, such that

$$g(\theta + G \cdot \delta) = 0. \tag{20}$$

Newton-Raphson and quasi-Newton methods can be and are defined explicitly for solving systems of equations. There are, of course, some practical differences in applying one of the methods to (19); for example,

(i) because g and G are not related to minimization, the procedures for proving and ensuring convergence are different, and

(ii) because (20) comes directly from (19) and not from a quadratic approximation, additional methods and variants are possible, such as nonsymmetric choice of G.

In (i) notice that the descent condition in Section **b** no longer holds, even at the solution.

The mathematical theory of nonlinear equations appears in a number of books: e.g., the text by Ortega and Rheinboldt (1970). A basic tool in proving convergence results is the *fixed-point theorem* (or contraction theorem or theorem of centers):

> *Theorem.* Suppose $\Gamma(\theta)$ takes an arbitrary real m-vector, θ, and returns a real m-vector as result. Then if there exists a scalar α, $0 < \alpha < 1$ such that for any θ and θ_*,
>
> $$||\Gamma(\theta) - \Gamma(\theta_*)|| \leqslant \alpha||\theta - \theta_*|| \tag{21}$$
>
> the sequence $\{\theta_j\}$ defined for arbitrary θ_1 by
>
> $$\theta_{j+1} = \Gamma(\theta_j)$$
>
> has a unique limit; i.e., $\theta_j \rightarrow \hat{\theta}$ regardless of θ_1.

An operator, Γ, satisfying (21) is called a *contraction mapping.*

Based on the idea of a contraction mapping and on related properties of mappings, convergence results can be derived for the Newton-Raphson, quasi-Newton and other methods. For a detailed discussion see Ortega and Rheinboldt (1970, Chapters 10 and 12), and Broyden, Dennis and More (1973).

Rate-of-convergence results can also be proved directly in terms of equations. The same references just mentioned prove, respectively, conditions for Newton-Raphson and quasi-Newton procedures. The continuity assumptions are similar to the minimization results, and one again obtains second-order convergence for Newton-Raphson and superlinear for quasi-Newton.

There exist a number of specialized methods for estimating parameters whose justification lies in statements of the following style.

> Given data Y, a model for Y involves K sets of parameters, say $\theta_{(1)}, \ldots, \theta_{(K)}$. If any K−1 of the sets are fixed, the optimum values of the Kth (by some criterion) are given by the solution to an easy problem. Therefore, we propose to compute optimum $\hat{\theta} = [\hat{\theta}_{(1)}, \ldots, \hat{\theta}_{(K)}]$ by cyclically fixing K-1 of the $\theta_{(j)}$ and solving for the Kth.

Variations exist, for example, in not strictly cycling through the $\theta_{(j)}$, but let us consider the simplest form. Examples of estimation by this procedure are the NIPALS procedure of Wold (1966, 1973), and the generalized linear models (Nelder and Wedderburn, 1972). "Easy" generally means operations of linear algebra or still simpler procedures.

The arguments in favor of such methods emphasize the ease of application; i.e., if handy procedures for solving the easy subproblems exist, then repeated application to solve the general problem requires relatively little extra programming. The user of a system or specialized language in which the development of new algorithms is more restricted or expensive than in most general programming languages will particularly appreciate this advantage.

On the other hand, there has been a tendency to use these ad hoc techniques without much consideration of their behavior as numerical methods, and without trying to compare them to more systematic computations.

To suggest some basis for systematic treatment, suppose we define $\Gamma_j(\theta)$ as the function that produces $\hat{\theta}_{(j)}$, given the current $\theta = [\theta_{(1)}, \ldots, \theta_{(K)}]$. Notice that we can broaden the definition to define Γ_j in nearly any way, without changing the essential argument to follow.

The application of K cyclic steps of this form produces a transformation generating a new set of estimates for all the components of θ. Formally, we define iterates as follows: $\theta_i = [\theta_{i,1}, \ldots, \theta_{i,k}]$ is generated from θ_{i-1}, if $i = rK + j$, $1 \leqslant j \leqslant K$, by the relationship

$$\theta_{i,j} = \Gamma_j(\theta_{i-1})$$

$$\theta_{i,L} = \theta_{i-1,L} \quad L \neq j. \tag{22}$$

Then every K^{th} estimate from (22) gives an estimate of the full parameter that we can write symbolically as

$$\theta_{i+K} = \Gamma(\theta_i)$$

or generally

$$\tilde{\theta} = \Gamma(\theta).$$

The properties of Γ can be derived from (22) and the properties of the individual Γ_j. In particular, the derivative of Γ, which is essential to many theoretical results, can be determined.

Suppose $\tilde{\theta} = (\tilde{\theta}_1, \tilde{\theta}_2, \ldots, \tilde{\theta}_K)$. Then (22) says that

$$\tilde{\theta}_j = \Gamma_j(\tilde{\theta}_1, \ldots, \tilde{\theta}_{j-1}, \theta_j, \ldots, \theta_K).$$

Applying the chain rule for composite functions (Fleming, 1965, p. 105) gives a recursive formula:

$$\partial\tilde{\theta}_j / \partial\theta_i = \sum_{r<j} (\partial\Gamma_j/\partial\theta_r(\partial\tilde{\theta}_r/\partial\theta_i) + \partial\Gamma_j/\partial\theta_i. \tag{23}$$

Assuming the individual Γ_j differentiable, then so is Γ, and its derivatives are given by (23).

Let us turn to a simple, but common, special case: *iterated weighted least squares.* Here we begin with a linear model and a fitting criterion to minimize a weighted sum of squared residuals, as in Section **5.h**.

In the commonest and simplest case, the weights, W, are defined in terms of the data and the regression coefficients, B. For example, robust linear regression methods are often of this form. Then one may write

$$\theta = [B, W]$$

and the iteration consists of two parts. First, given W, one computes $\tilde{B} = \Gamma_1(W)$ by solving the linear regression in Equation (46) of Section **5.h**. Second, one computes new weights, $\tilde{W}$, from $\tilde{B}$ by a known function, $\Gamma_2(\tilde{B})$. Hence $\tilde{\theta}$ is defined.

In order to verify fixed-point results derivatives of Γ_1 and Γ_2 are needed. For Γ_2 this is generally straightforward. Derivatives of the linear least-squares operator are discussed by Golub and Pereyra (1973).

Therefore, by routine if somewhat laborious calculations, the functional derivative of the iteration can be computed for any specific model structure. The relevance of such a calculation is that the powerful theory of iterative methods can be applied. In particular, if $\hat{\theta}$ is a solution to (1), and one can compute $\partial\Gamma/\partial\theta$ at $\hat{\theta}$, one can apply the following test (Ortega and Rheinboldt, 1970, p. 300): if $\Gamma(\theta)$ is differentiable at $\hat{\theta}$ and

$$|\lambda_1| < 1$$

where $|\lambda_1|$ is the maximum absolute value of the eigenvalues of $\partial\Gamma/\partial\theta$ at $\hat{\theta}$, then the iteration $\theta_{i+1} = \Gamma(\theta_i)$ converges to $\hat{\theta}$ if θ_1 is in a sufficiently small neighborhood of $\hat{\theta}$.

Unfortunately, few of the fixed-point methods have included such results, which would clarify and support the methods' use.

i. Constrained Optimization

The elements of the parameter, θ, have been assumed free to take on any real values. Should this not be the case, the problem becomes one of *constrained optimization* or *nonlinear programming.* The requirement is usually expressed as a set of (nonlinear) equality and/or inequality constraints; namely, to find θ to minimize $F(\theta)$ such that, for specified functions, c_i,

$$c_i(\theta) \geqslant 0 \quad i = 1,\dots,k$$

$$c_i(\theta) = 0 \quad i = k+1,\dots,k+r. \tag{24}$$

In data analysis, the $c_i(\theta)$ may depend upon the data, Y, as well as the parameters. The case that $k = 0$, with only equality constraints, is mathematically and computationally special.

For the general inequality constraints, the most actively studied methods are the *sequential unconstrained minimization techniques* (SUMT) in which a sequence of functions, say $\Phi_{(1)},\Phi_{(2)},\dots$, are defined so that the unconstrained minimum of $\Phi_{(j)}$ will approach, as $j \to \infty$, the constrained minimum desired. Both the inspiration and the mathematical support for these methods come in part from the result known as the *Kuhn-Tucker Theorem.* A function of $(m + k + r)$ variables, $\Phi(\theta,\lambda)$, is defined by

$$\Phi(\theta,\lambda) = F(\theta) - \sum_{i=1}^{k+r} \lambda_i c_i(\theta). \tag{25}$$

Then under suitable conditions on $F(\theta)$ and $c_i(\theta)$, the constrained optimum $\hat{\theta}$, corresponds to a point $(\hat{\theta},\hat{\lambda})$ at which

$$\partial\Phi/\partial\theta = 0$$

$$\partial\Phi/\partial\lambda = 0. \tag{26}$$

See Mangasarian (1969) for a readable discussion. The elements of λ are called the Lagrange multipliers, and the equality constraint case is the classical Lagrangian result (Fleming, 1965, 129-30).

In a general sense, the computational methods attempt to find the stationary point (26) as the limit of a sequence of minima. There are many variants, originally using *loss* or *barrier* functions. A loss function, $L(x)$, is zero if $x \geqslant 0$ and increases to infinity as $x \to -\infty$. One form of sequential minimization is then to define

$$\Phi_{(j)}(\theta) = F(\theta) + \delta_j^{-1} \sum_i L(c_i(\theta)) \tag{27}$$

choose a sequence of δ_j tending to zero, and successively minimize $\Phi_{(j)}$.

A barrier function, $B(x)$, tends to infinity as x tends to zero from above and decreases as $x \to \infty$. These may be applied in sequential minimization by defining

$$\Phi_{(j)}(\theta) = F(\theta) + \delta_j \sum_i B(c_i(\theta)). \tag{28}$$

The methods based on (27) and (28) are called external and internal methods because the $\hat{\theta}_{(j)}$ which minimizes $\Phi_{(j)}(\theta)$ is outside or inside the feasible region, respectively. It is also possible to mix the two by adding the terms in both (27) and (28) to $F(\theta)$. One then assigns a term $L(c_i(\theta))$ if $c_i(\theta) < 0$ and $B(c_i(\theta))$ otherwise.

Interest in sequential minimization techniques was stimulated in part by the book of Fiacco and McCormick (1968). Considerable use was subsequently made of fairly simple loss and barrier functions. The choice of $L(x) = x^2$ and $B(x) = \log(x)$ is common; one must then choose some δ_j values, say, $\delta_j = \delta_0^{-j}$. After a certain number of δ_j and corresponding $\hat{\theta}_{(j)}$ have been computed, one may attempt to fit a univariate curve in δ to the elements of θ and extrapolate to $\delta = 0$ (Fiacco and McCormick, 1968). On balance, the mixed penalty function is probably best for home-made algorithms. However, there are many practical and theoretical difficulties. A major one is that the derived unconstrained problems become increasingly ill-conditioned as $\delta_j \to 0$. Two general alternative approaches are to look for solutions to the Lagrangian problem, (25) and (26), directly, and to construct more attractive penalty functions. Within the latter approach, one method has been to construct single unconstrained functions whose minima are solutions to the constrained problem. Work by Fletcher (1973) and Conn and Pietrzykowski (1977) is promising, but the former requires derivatives of the original problem to construct the unconstrained function and the latter involves the minimization of a a non-differentiable function.

The Lagrangian methods are perhaps the most promising at present. Usually, a sequence of problems are solved, in terms of θ and λ in (25), which tend in the limit to a solution to (26). One technique, described by Garcia-Palomares and Mangasarian (1976), solves a sequence of quadratic programming problems. Algorithms for the latter are available (see the end of this Section). For a survey of related methods, see Wright (1977).

Many other methods for constrained optimization exist. *Projection* methods begin with a minimization technique applied to $F(\theta)$ alone, but modify the unconstrained step in such a way as to guarantee that constraints are not violated. For example, if the constraints were linear equality constraints, one could project any step direction (e.g., $-H_i \cdot g_i$ in a quasi-Newton method) onto the surface satisfying the constraints. Extensions of the technique exist for general constrained optimization. One may approximate nonlinear constraints by local linear Taylor series. The inequality problem is transformed locally to equality by knowing which constraints are currently active. Methods of this general type have been presented by Zoutendijk (1960), Greenstadt (1966), Murtagh and Sargent (1969), and others.

The general search methods have also been extended to constrained optimization. A method based on the simplex search of Spendley et al (1962) was developed by Box (1965) and exists in a published FORTRAN algorithm by Richardson and Kuester (1973). As in unconstrained optimization, the search methods are less attractive when high accuracy is needed or when the number of parameters is large, but may have advantages on functions that are not smooth.

As mentioned, the case of equality constraints is specialized. The penalty function approach can be carried over fairly directly; for example, by using a loss, $L_{(j)}(x)$, that tends, as $j \rightarrow \infty$, to zero at $x = 0$ and to infinity elsewhere. Numerical instabilities associated with this approach, as in the general constrained problem, have stimulated methods to solve the Lagrange equations more directly (Powell, 1969; Haarhoff and Buys, 1970). In a series of papers Fletcher (1970a; 1972; 1973) and Fletcher and Lill (1970) developed an algorithm that minimized a single function (involving $F(\theta)$, $c(\theta)$ and the derivatives of $c(\theta)$). Fletcher (1972) also considers the special case of linear constraints. At the time of writing, no published algorithms exist for these methods.

The special cases of *linear* and *quadratic* programming have been extensively treated in economics, engineering, and other nonstatistical fields. For linear programming the standard method is the simplex procedure for searching among the vertices of the region in the parameter space, within which the constraints are satisfied. Bartels and Golub (1969) give an ALGOL60 algorithm using triangular matrix decompositions, and Gill and Murray (1970) describe verbally an algorithm using orthogonal decompositions. See also the Appendix. Many computer manufacturers or software houses also provide linear programming systems. Those interested in further discussion of linear programming may consult Dantzig (1963), Spivey and Thrall (1970) or, for a more computational approach, Luenberger (1974).

The case of *quadratic programming,* in which the objective function is quadratic and the constraints linear, may also be treated by a form of simplex method (Beale, 1967; Rusin, 1971). The special case of constrained linear least-squares estimation arises most frequently in data analysis. When the constraints are equalities, the standard decompositions can be modified to include the constraints directly. Algorithms in ALGOL60 are given in Björck (1968) and Björck and Golub (1967).

The available general algorithms for general constrained optimization do not represent fully the latest methods. There is some confidence among specialists in this area, however, that sufficient knowledge has now been accumulated that, within the near future, high-quality algorithms will be more generally available. Some of the program libraries contain algorithms (see the Appendix). For actual published algorithms, the book by Himmelblau (1972) gives some FORTRAN programs (not, unfortunately, portable).

j. Summary and Recommendations

A good model-fitting system should provide several of the better numerical methods. First, for general optimization:

1. a Newton-Raphson method, with some built-in procedure for stabilizing the method;
2. one or more quasi-Newton methods, preferably one of the later versions, incorporating some modification of the original line search and a factored form of the Hessian;
3. a search routine, such as Nelder and Mead (1965);
4. a conjugate gradient method.

The choice among the four possibilities is essentially as follows. Search routines should be restricted to problems in which the number of parameters is not too large and high accuracy is not essential, and are particularly appropriate for functions which are not smooth. Conjugate gradient methods are useful mainly when the number of parameters is too large to work with the matrices entering into the first two methods. The quasi-Newton method will likely be the first choice for most problems, unless the second derivatives are easy and cheap to evaluate.

One or more specialized routines for nonlinear least squares are also desirable:

5. a stabilized Gauss-Newton method.

At the present, no single method for constrained optimization can be clearly recommended, and the available algorithms are not particularly satisfactory. The methods of approximating the Lagrangian function are promising, but older methods such as penalty-barrier and projection methods are more commonly available and may perform adequately for many problems.

The availability of auxiliary information contributes to the statistical usefulness as well as the convenience of nonlinear fitting. The user should be supplied with the asymptotic variance estimates if available, but as often as possible one should go beyond these to simulation and other model descriptions.

As this is written, the available algorithms for nonlinear fitting are significantly inferior to the best methods currently known, particularly for those who do not have access to the program libraries cited in the Appendix. A cautiously optimistic view is that the theoretical work of the late 1960s and 1970s has clarified many of the problems, so that high-quality algorithms reflecting the advances may be generally available by 1980 or so.

A few more fundamental problems remain. One is that users must provide the definition of a model, in the form of a subprogram to compute $F(\theta)$ and possibly its derivatives, or the equivalent in other forms of model-fitting. The tendency to make mistakes in this formulation is one of the major difficulties in nonlinear fitting. We need a general facility to allow simpler definition of models. Existing work on analytic differentiation provides one component of the job, but systems that co-ordinate powerful analytic differentiation techniques with optimization, particularly in a portable way, do not yet exist. Another difficulty is that the performance of some algorithms is much affected by reformulating the model in some way such as rescaling or transforming the parameters. We still know very little about how to choose the right form of the model. Initial estimates are also an obviously important factor, always for the parameters and sometimes for other quantities as well (see Section c). Heuristic knowledge about such questions needs to be accumulated for subclasses of nonlinear models and made available to users of a fitting system.

CHAPTER SEVEN

Simulation of Random Processes

A major contact point between statistical and computational theory is the attempt to produce values by a computer that behave as if they were the result of some random process. A reliable source of such values would be of great benefit in computation, as well as providing statisticians and others with test data for models involving randomness.

a. The Concept of Randomness

The attempt to generate values with random properties raises a number of questions, both practical and philosophical. Many of the latter involve asking what randomness means and how one could decide whether an observed process was random or not. The simplest precise statement of the problem would be: What criterion will determine whether an infinite sequence of bits, $\{b_i\}$, is equivalent to a sequence of independent random observations, $\{B_i\}$, with

$$\begin{aligned} \text{Probability}(B_i = 1) &= p \\ &= 1 - \text{Probability}(B_i = 0)? \end{aligned}$$

One is looking, in general, for some mathematical, deterministic property which $\{b_i\}$ must possess, in order to be random. Von Mises (1957) gave a famous but not very precise criterion: any chosen infinite subsequence of the b_i should also have a limiting frequency of p for the number of 1's, provided that the values of the b_i were not used in the selection. A more precise and more general definition would be as follows: Let $i_k(\beta_1, \ldots, \beta_{k-1})$ be an arbitrary effectively computable function of $k-1$ binary arguments yielding positive integer values, for $k = 1,2, \cdots$. Then a subsequence $\{\beta_j\}$ of $\{b_i\}$ can be defined by recursion:

$$\beta_1 = b_1$$

$$\beta_j = b_i \text{ with } i = \sum_{k=1}^{j} i_k(\beta_1, \ldots, \beta_{k-1}) \tag{1}$$

Then $\{b_i\}$ is said to be *random* if every choice of the functions, i_j, leads to a sequence, $\{\beta_j\}$, with limiting frequencies the same as $\{b_i\}$.

This simply makes precise the intuitive notion that no calculations based only on the sample taken so far can change the odds that a specific future observation (e.g., the one coming $i_k(\beta_1, \ldots, \beta_{k-1})$ ahead) will be a 0 or 1. The term *effectively computable* is used here in its technical sense (Minsky, 1967, Chapter 5), but the intuitive interpretation is again clear: we assume an algorithm could be written for i_k. Such a definition seems to conform to introspection about what randomness means and is consistent with the mathematical, axiomatic formulation of probability. (For a similar definition, based on a more complicated and extensive argument, see Knuth (1969, Section 3.5).)

However, the reader should not be misled by the computational flavor of the definition. Randomness, so defined, is not operationally verifiable. In other words, there cannot be a single algorithm, Q, such that given any sequence $\{b_i\}$, Q will always terminate after looking at a finite number of elements of $\{b_i\}$ and will correctly decide whether the sequence is random. Intuitively, if there were such a Q, we could write an algorithm B that always generated b_i so as to keep Q from deciding that $\{b_i\}$ was nonrandom (the definition of B obviously depending on the definition of Q).

On the other hand, no effective algorithm, B, can generate a random sequence by this definition. For if B is an algorithm, the algorithm for the j-th 1 bit generated by B is also computable for any finite j. In particular, we can effectively select the subsequence $\{1,1,1, \cdots\}$ from $\{b_j\}$, and therefore B is not random. Hence our assumption of the existence of Q must be wrong.

These abstract discussions have the practical point that one should not talk about showing that a generated sequence is random. If the generator is an algorithm, we know the sequence is not random. If the generating mechanism is nonalgorithmic (for example, generators based on particle emissions), we know that the question of randomness is undecidable in general.

The sensible question to ask, therefore, is not "Is this sequence random?" but "Will this sequence behave as if it were random, *when applied to my problem?*" For problems of practical importance, a definitive answer is rarely possible, but there exist many theoretical and empirical analyses of specific properties for generators. Using these and some judicious preliminary testing, the user can often obtain psychological comfort about the validity of his simulation results.

The consideration of simulation algorithms breaks down into three stages: the generation of uniform pseudorandom numbers, the generation of other distributions and random processes, and the application of the generated numbers to computing problems (i.e., Monte Carlo procedures).

b. Pseudorandom Uniforms

The computational problem can be stated several ways, of which the following is perhaps the least confusing: A uniform pseudorandom generator is a mechanism for generating a sequence of fractions $u_1, u_2, \ldots, u_N$, $0 < u_i < 1$, such that $\{u_i\}$ will replace random observations from the uniform distribution on (0,1) and give satisfactory answers for simulation purposes. Generally, u_i is defined modulo some integer P; i.e., the fractions are computed by

$$u_i = r_i / P$$

from the sequence $\{r_i\}$ of corresponding integers. The r_i are often called *pseudorandom integers,* but this term needs to be used with caution, as the r_i from many generators have serious deficiencies. If the sequence repeats after N numbers, N is called the *period* of the generator.

The techniques which are currently in use and seem competitive for some problems are:

1. congruential generators;
2. shift-register generators;
3. nonalgorithmic generators;
4. post-processing applied to one or more basic generators.

Sections **c** to **e** discuss examples of these generators. There are, of course, many other algorithms chiefly notable as historical curiosities or monstrosities. The book by Jansson (1966) gives a good historical review. Generally Type 4 generators, which were usually developed to compensate for known nonrandom behavior of the simpler generators, are recommended for most applications.

c. Congruential Generators

The linear congruence relationship defines $\{r_i\}$ by

$$r_i = \mathrm{mod}(r_{i-1} \times \lambda + \alpha, P) \tag{2}$$

where mod(i, j) is the remainder or *residue* when i is divided by j, as integers. This mechanism has been very widely used to generate pseudorandom fractions. The precise sequence generated by (2) depends on the choice of P, λ, α and r_1. To make remaindering simple, P tends to be chosen to be 2^i for some i. When $\alpha = 0$, the generator is called *multiplicative congruential,* otherwise it is called *mixed congruential.*

The simple relationship in (2) can be studied by number theoretic methods, giving some useful specific formulas for the period and other properties. A number of such results have been published. Marsaglia (1972) states a number of results based on the use of the *fundamental sequence* $\{s_i\}$ defined for given P and λ as

$$s_1 = 0$$

$$s_i = \mathrm{mod}(s_{i-1} \times \lambda + 1, P); \tag{3}$$

i.e., by taking $\alpha = 1$ and $r_1 = 0$. One can show directly that (2) is related to (3) by

$$r_i = \mathrm{mod}(v \times s_i + r_1, P)$$

where $v = \mathrm{mod}(r_1 \times (\lambda - 1) + \alpha, P)$. This does not quite imply that the properties of (2) are completely defined by (3). However, many of the number-theoretic results can be derived directly for the fundamental sequence and then applied to the general case.

A simple formula of frequent use expresses r_{i+j} in terms of r_i:

$$r_{i+j} = \mathrm{mod}(\lambda_j \times r_i + \mu_j, P) \tag{4}$$

with

$$\lambda_j = \lambda^j$$

$$\mu_j = (1 + \lambda + \cdots + \lambda^{j-1})\mu \tag{5}$$

$$\doteq \frac{(\lambda^j - 1)}{(\lambda - 1)} \cdot \mu .$$

This result follows directly from the definition of the sequence and is applied in studies of multiple values.

Three topics have been treated successfully for congruential generators by number-theoretic methods: the *period* of the generators, some results on *moments* and the distribution of derived *multidimensional* variables. The generator (2) has a period that is difficult to derive for arbitrary P but is much simpler in the case $P = 2^i$. The maximal period is particularly easy to derive and is summarized as follows:

> **Theorem.** The maximal period of the mixed congruential (2) is 2^i and is obtained if and only if
>
> (i) $\mathrm{mod}(\lambda,4) = 1$
>
> (ii) α is odd.
>
> The maximal period of the multiplicative congruential generator is 2^{i-2} and is obtained if and only if
>
> (iii) $\mathrm{mod}(\lambda,8) = 3$ or 5
>
> (iv) r_1 is odd.

The proof of these results is fairly straightforward from classical results in number theory. The interested reader is referred to Jansson (1966, Chapter 3) or Knuth (1969, Section 3.2.1.2). Marsaglia (1972, p. 262) gives an expression for the general period, but presumably one would always choose the maximal case.

Marsaglia (1972) extended the notion of period by showing that the sequence, $\{s_i\}$, obeys the relation

$$s_{t+i} = \mathrm{mod}(s_{t+1} + s_i, P) \tag{6}$$

for t the smallest integer such that $\lambda^t - 1$ is a multiple of P. In number theory t is called the multiplicative order of λ, mod P. The essential property of t is that

$$\mathrm{mod}(\lambda^j \lambda^t, P) = \mathrm{mod}(\lambda^j, P) \tag{7}$$

for $j \geqslant 0$. It is easy to show from (3) that

$$\begin{aligned} s_{t+i} &= 1 + \lambda + \cdots + \lambda^{t+i-2} \\ &= s_{t+1} + \lambda^t + \lambda\lambda^t + \cdots + \lambda^{i-2}\lambda^t \end{aligned}$$

from which it follows, using (7), that

$$\mathrm{mod}(s_{t+i}, P) = \mathrm{mod}(s_{t+1} + s_i, P).$$

Equation (6) says that the sequence, $\{s_i\}$, consists of a block of t numbers followed by another block consisting of translates of the first, mod P, followed by another translated block, and so forth. Marsaglia argued that the *effective period* of the generator is t. One can not quite reduce all the analysis to the simpler analysis of a block of t numbers, since the statistical properties of the block will change when it is translated and then residues are taken. However, if one wished to make the effective period as large as possible and the period also as large as possible, it is necessary, in the case of a mixed congruential with $P = 2^i$, that

$$\lambda = 1 + k \times 4$$

where k is an *odd* integer.

Some of the simpler moments of $\{r_i\}$ can be calculated directly from a knowledge of the set of numbers generated. The mean and variance of the full period sequence of the mixed congruential are easily derived, since the generated set is $0, 1, \cdots, P-1$. Therefore

$$\sum_{i=1}^{P} r_i = P(P-1)/2$$

$$\sum r_i^2 = P(P-1)(2P-1)/6.$$

From this result it follows that the mean and variance of the full period sequence are $(P-1)/2$ and $(P^2-1)/12$ respectively. The floating point values, r_i/P have

$$\text{mean} = \frac{P-1}{P} \cdot \frac{1}{2}$$

$$\text{variance} = \frac{P^2-1}{P^2} \cdot \frac{1}{12}$$

which for P large differ trivially from the values, 1/2 and 1/12, for the uniform distribution.

The serial correlations depend on the specific generator, and no closed form expression exists. However, Jansson (1966, Chapter 6) develops some fairly simple algorithms to determine the serial correlations from recursions. Numerical results obtained by Jansson for a specific set of generators with $P = 2^{35}$ showed serial correlations that were significantly closer to zero than statistical theory would have predicted. Unfortunately, the class of λ chosen was of the form $\lambda = 2^j + 1$, a choice which the multiple-value tests discussed later in this Section show to be particularly poor. Because there were no global bounds given independent of λ, the generality of Jansson's results is not clear.

Similar in concept to serial tests are studies of the *spectrum* or Fourier transform of the generators. Indeed, spectra and serial correlations are frequently competitors in statistical analysis. (For an introduction to the use of spectra, see Jenkins and Watts (1968) and Section **4.h**.) If we consider the sequence of integers, $\{r_i\}$, we can take the finite Fourier transform of k successive values, producing a transform, $\phi(s_1, s_2, \ldots, s_k)$:

$$\phi(s_1, \cdots, s_k) = \frac{1}{P} \sum_j \exp(-\frac{2\pi i}{P}(s_1 r_j + \cdots + s_k r_{j+k-1}))$$

This immediately simplifies because the mixed congruential generator implies

$$r_{i+j} = \mathrm{mod}(\lambda_j r_i + \mu_j, P) \tag{8}$$

where λ_j and μ_j are defined as

$$\lambda_j = \lambda^j$$

$$\mu_j = \frac{(\lambda^j - 1)}{(\lambda - 1)} \mu$$

In fact, this is exactly the relation used to derive the serial correlation formulas. As a result of (8) the Fourier transform can be expressed neatly as follows:

$$\phi(s_1, \cdots, s_k) = \begin{cases} \exp(-\frac{2\, i\mu s_*}{P}) \\ 0 \end{cases}$$

where

$$s_* = \frac{\mu}{\lambda - 1} \sum s_j(\lambda^{j-1} - 1)$$

and $\phi = 0$ unless $\sum s_j \lambda^{j-1}$ is a multiple of P. As with Jansson's results, we can now compute spectra for specific generators. The ideal would be for ϕ to be nearly zero except when $s_1 = s_2 \cdots = s_k = 0$, if the r_j are to behave as if they were independently distributed. Thus one can compare specific generators. Coveyou and Macpherson (1967) developed such tests and attempted to summarize the spectrum by a *wave number,* ν_k. Notice that $\phi(s) = 0$ unless s satisfies

$$\mathrm{mod}\left(\sum s_j \lambda^{j-1}, P\right) = 0. \tag{9}$$

Coveyou and Macpherson defined ν_k^2 as the minimum value of $\sum s_i^2$ for which (9) holds. The larger ν_k the better the generator, other things being equal. Equation (9) is a quadrature in k integers; a number theoretic result called Hermite's theorem bounds the resulting values of ν_k by

$$\nu_k \leqslant c_k P^{1/k}$$

with $c_k < (4/3)^{(k-1)/2}$. Thus one can compare the performance of a generator to a best possible bound. Coveyou and Macpherson use these results to evaluate a number of λ values and to suggest some good choices for this criterion.

The studies of serial correlations and Fourier transforms provide evidence on the distribution of multiple values from congruential generators. A different approach to this problem is to consider the effect of taking integer linear combinations of the k-dimensional points $(r_i, r_{i+1}, \ldots, r_k)$. The set of all points that can be generated by such linear relations is called the *k-lattice* of the generator by Marsaglia (1972), who takes as an optimum the criterion that the lattice should be complete; i.e., the set of all points generated by the k coordinate axes, $(0, \cdots, 0, 1, 0, \cdots, 0)$. It happens again that lattices for congruential generators can be analyzed very simply.

Suppose we generate k-dimensional pseudorandom uniforms by the natural scheme:

$$P_1 = (p_1, \ldots, p_k),\ P_2 = (p_{k+1}, \cdots, p_{2k}),\ \cdots,$$

so that $P_j = (p_{(j-1)k+1}, \cdots, p_{jk})$. Clearly, this is a type of simulation that will arise in practice, either directly because we want k-dimensional uniforms or indirectly because we want a derived variable that requires k uniforms for each realization. The questions to be asked then are how far the realized P_j differ from the uniform distribution in the k-dimensional unit cube, how much such deviations affect our use of the numbers, and what can be done to alleviate problems.

In fact, the general point in k dimensions, say P, is always related to a fixed origin, P_0, by

$$P - P_0 = i_1 P_1 + \sum_{j=2}^{k} i_j m \delta_j \tag{10}$$

where $P_1 = (1, \lambda, \lambda^2, \ldots, \lambda^{k-1})$, δ_j is the vector with $\delta_j[j] = 1$, and $\delta_j[i] = 0$ for $i \neq j$. Therefore the integer linear combinations of points P are all spanned by P_1 and $\{m\delta_j\}$. This means that the unit cell (smallest parallelotope) in this lattice has volume

$$D = \det \begin{bmatrix} 1 & \lambda & \lambda^{k-1} \\ 0 & m & 0 \\ & \cdots & \\ 0 & \cdots & m \end{bmatrix} \qquad (11)$$

$$= m^{k-1}$$

Intuitively, one would like a generator to produce integer points "uniformly" in k-space, which may be interpreted as implying that the lattice above should be spanned by the unit axes, $\delta_1, \ldots, \delta_k$. In this ideal case the unit volume would be $D = 1$. Marsaglia interprets (11) as a measure of the nonuniformity of congruential generators in this respect. While λ has no effect on D, one can attempt numerically to find the basis for the lattice that minimizes some measure of its departure from the "perfect" $\delta_1, \ldots, \delta_k$. Marsaglia (1972) describes an algorithm for this and selects some good λ values based on the results.

These are some of the main theoretical or semi-theoretical results. Notice that they relate to the full set of values generated. No number-theoretic results are available for shorter sequences, although these represent the practical use of the generators. Worse, most of the sequential results refer to the set of points $(r_i, r_{i+1}, \ldots, r_{i+k-1})$, for *all* i, a set that will not even be generated by one pass through the entire generator, unless we use k successive starting points. Therefore, one may feel that the number-theoretic results are at best suggestive, rather than definitive.

Many empirical studies also exist. These usually generate sequences of u_i, compute summaries and compare the results to the known distribution of the summaries for samples from the uniform distribution. The gist of the empirical tests has been that some multipliers, at least, seem to produce bad distributions in k-space (Maclaren and Marsaglia, 1965; R. P. Chambers, 1967), but that the better congruential generators are not strikingly nonuniform in most empirical tests (Lewis et al, 1969).

Despite categorical statements by various authors, then, the picture is confused. Roughly, number-theoretic results are disquieting, at least for the generation of multidimensional variables, but so far specific statements relating these results to practical use are lacking, and empirical results on good congruential generators, at least, are inconclusive. More relevant theory or more incisive empiricism may well clarify the nature and seriousness of nonuniformities.

Practical prescriptions will vary and there has been an abundance of strongly worded discussion. The author's feeling is that present knowledge is inadequate to say when and to what extent use of congruential generators will give significantly misleading results in practice. Nevertheless, there is enough indirect evidence that a cautious user will want to take some precautions. At the moment, this will usually mean applying one of the postprocessing techniques of Section **e**. These do not guarantee to solve problems, but they are attractive intuitively and the additional cost involved will not be a major factor in most applications.

d. Other Basic Generators

Most procedures suggested for generating pseudorandom fractions are only historical curiosities now. Jansson (1966) and Knuth (1969, Section 3.2.2) summarize early efforts. Two generating schemes still in current use deserve comment. One is the class of mechanisms that may be called *natural generators* in that they rely upon the asserted randomness of some physical phenomenon to induce apparent randomness in the output. The other class is the *shift-register generators* and related procedures using certain bit-by-bit congruences.

Natural generators usually rely on recording some process, such as radioactive decay, that physical theory postulates to occur in a random manner, following some specific distribution. There are several approaches, and many physical devices may be candidates. The physical-theory model may predict noise as a continuous variable with, for example, a normal distribution, or it may predict the occurrence of discrete events according to some probability law. If an on-line device were to be developed, one would want to convert the input from the generator into a standard form, say as parallel input of l bits, supposed to be identically and independently distributed Bernoulli processes, giving 0 or 1 with equal probability on each cycle. Such a generator could be produced by hardware or software processing of other signals such as normal noise (Murry, 1970). The catch, in any case, is that our belief in the numbers produced requires faith in both the validity of the physical theory describing the generator and in the stability of the device itself through time. While devices exist that are claimed to be satisfactory on both counts, they have not been widely adopted so far in general simulation applications.

A different use of natural generators is to produce a large pseudorandom data set that is then treated as a finite sequence from which numbers are drawn for simulation. This does not provide an unlimited sequence of numbers and requires storing a large amount of data. However, it is possible to subject the numbers to considerable testing and postprocessing in an attempt to verify or improve their performance in the tested situations. The numbers can then be copied and transmitted to other computer installations, which may not have built-in generators. The best known data set of this form was developed by the RAND Corporation (1955) and is still used fairly widely.

Notice that built-in natural generators are random rather than pseudorandom in the one sense that the user cannot control the output so as to reproduce a specific simulation. Whatever one thinks of this philosophically, it does make debugging and verification of Monte Carlo results less convenient. For this reason one might wish to use a built-in generator to produce a data set of numbers, even if the quality of the natural generator was unimpeachable.

The *shift-register generators* arose in studies of coding, information theory, and communication, subsequently being proposed for generating pseudorandom bits. The book by Golomb (1967) gives a general account of shift-register generators in communications. The essential notion is that of a linear recurrence generating a sequence of bits $\{b_i\}$:

$$\begin{aligned} b_i &= \mathrm{mod}(c_1 b_{i-1} + c_2 b_{i-2} + \cdots + c_d b_{i-d}, 2) \\ &= \mathrm{mod}\left(\sum_{j=1}^{d} c_j b_{i-j}, 2\right) \end{aligned} \tag{12}$$

where c_i are all 0 or 1. Clearly one needs to keep d successive bits available to generate the sequence. We can imagine these as forming unsigned integers of length d:

$$\begin{aligned} r_1 &= b_1 b_2 \cdots b_d \\ r_2 &= b_2 b_3 \cdots b_{d+1} \\ &\cdot \\ &\cdot \end{aligned}$$

or, in general,

$$\begin{aligned} r_i &= b_i\, b_{i+1} \cdots b_{i+d-1} \\ &= \sum_{j=0}^{d-1} 2^{d-j-1}\, b_{i+j} . \end{aligned} \tag{13}$$

Like congruential generators, shift-register generators have properties directly deducible from the definition (13). These properties may be weighed for or against the use of the generators in practice.

The maximal period of the linear generator in (12) is $2^d - 1$ and is obtained if and only if the polynomial

$$P(x) = 1 + \sum_{j=1}^{d} c_j x^j$$

is *primitive* when x has values 0 or 1. See Knuth (1969, Section 4.6) for some discussion of primitive polynomials. The case that has aroused most interest is that of primitive trinomials, which can be written

$$PT(x) = x^p + x^q + 1$$

for integers $p > q \geqslant 1$. These result in particularly simple algorithms to generate the sequence $\{r_i\}$; see, for example, Lewis and Payne (1973).

Whenever the generator has the maximal period, it follows that all the possible d-bit integers r_i, except 0, are produced provided $r_1 \neq 0$. Then it follows that mean and variance of the r_i, over the full period, are nearly that expected if the numbers were truly uniformly random. Tausworthe (1965) also shows that the autocorrelations are very small over the whole period for small lags.

Nothing so far suggests any strong advantages to shift-register over congruential generators. The feature most often cited is the distribution of k successive elements. Specifically, suppose we generate L-bit integers by chopping up the sequence $\{b_i\}$ in batches of

$$\begin{aligned} s_1 &= b_1\, b_2 \cdots b_L \\ &\;\;\vdots \\ s_j &= b_{(j-1)L+1} \cdots b_{jL} \end{aligned} \tag{14}$$

This is clearly a generalization of (13). It turns out that we should arrange that L and $2^d - 1$ have no common factors. As in the study of congruential generators, one considers the sequence of points

$$P_i = (s_i, s_{i+1}, \ldots, s_{i+k-1})$$

in k dimensions. Tausworthe (1965) showed that, for $kL \leqslant d$ the sequence $\{P_i\}$ is uniformly distributed over the k-dimensional cube. This is interpreted as saying that the generator gives k-dimensional uniformity, in contrast to the results for congruential generators.

Similar complaints can be made about this result to those mentioned for congruential generators. The theory only applies to the full sequence; worse, its last result uses all possible P_i, which are not generated by any single cycle. Tootill, Robinson and Adams (1971) did some more realistic, but painful, analysis using the "runs" test, i.e., the number of times r successive integers go monotonically up or down. They showed that Tausworthe's results did *not* guarantee good performance on this test and went on to find some reasonably acceptable generators. Marsaglia (1972) asserts that shift-register generators have bad lattice properties.

e. Modifying Generators

Several techniques have been proposed for postprocessing the output from one or more basic generators to remove bad features. Of note here are:

(i) shuffling;

(ii) pseudorandom replacement;

(iii) exclusive-or mixture.

Briefly these work as follows. Let $\{r_i\}$ come from a basic generator as above. Shuffling takes a block of M of the r_i and permutes them in an array, by some permutation phi, of (1,2,...,M). Thus, the j-th element of the block will be moved to position $\phi(j)$. Shuffling requires little extra work, although space for M values must be provided. The method requires choosing the permutation, ϕ. One possibility would be to take ϕ to be a pseudorandom permutation, and probably to regenerate ϕ every so often. A more specific motivation is to use ϕ to break up k successive values for various small values of k. Intuitively, one hopes that this may remove some of the k-dimensional regularities. This has been verified empirically in some studies of congruential generators. The extensive simulations at Princeton described in Andrews et al. (1973) used a shuffled congruential with a random permutation and $M = 500$.

Replacement generators require two generators, say $\{u_i\}$ and $\{u_i'\}$. Again a block of M numbers is retained, initially set to $u_1,...,u_M$. Then subsequently one uses u_i' to select an index j on the range 1 to M:

$$j = \mathbf{ceil}(u_i' \times M)$$

where **ceil** is the smallest integer at least as large as its argument. The *current* contents of the jth element is put out as the value of the generator and is then replaced by u_i. This method is originally due to Maclaren and Marsaglia (1965). See also Knuth (1969, p. 30).

A third method also uses two generators, but in their integer form: say $\{r_i\}$ and $\{r_i'\}$. The ith number generated is then just

$$\mathrm{XOR}(r_i, r_i')$$

where XOR is the bit-wise exclusive-or (the result has a 1 bit in a position if *exactly* one of r_i, r_i' has a 1 bit there). The success of this procedure depends, it seems, on $\{r_i\}$ and $\{r_i'\}$ being very different; e.g., a congruential and a shift-register generator. Marsaglia and others at McGill University developed a package of generators for the IBM/360/370 computers using just this combination. See also Section **j** and Gross (1976).

None of these procedures is guaranteed to remove all serious defects in the use of a basic generator. As we noted, this goal is not operationally attainable when a deterministic algorithm is used. The best one hopes is that some of the more blatant problems previously discovered will be alleviated. Method (i) essentially rearranges the terms in a sequence, and retains the full period properties of $\{u_i\}$, except of course the sequential properties. The same is nearly true of (ii), but the pseudorandom replacement alters the exact set of values produced over the full period of $\{u_i\}$. The properties of (iii) in this sense are harder to assess, particularly since the two basic sequences should differ in period.

In summary, basic generators of the congruential or shift-register types have a number of desirable properties, at least when the specific generator is chosen on the basis of studies such as Coveyou and Macpherson (1967), Marsaglia (1972) or Tootill et al. (1971). There remain some disquieting problems, particularly with multidimensional use of the generators, and the good properties are often not directly relevant to practical use. The postprocessing methods offer inexpensive devices that ought to improve at least some of the bad features. For applications needing accurate simulations, the extra cost is a moderate premium for the increased security provided.

The entire situation is still imprecise, however. We must wait and hope for more specific results on the errors introduced into simulations by treating various generators as random numbers.

f. Derived Distributions: General Methods

In most installations, pseudorandom integers and fractions are the basic material for simulation and other distributions and processes are generated from these. Natural generators might be developed to simulate directly other distributions such as the normal or Poisson, but this has rarely been done.

Several general techniques exist for generating arbitrary distributions. Specialized techniques take advantage of known properties of specific distributions. Some nonindependent random processes, such as order statistics and Markov processes, can also be given special treatment.

One rather trivial general method uses the *inverse distribution function* $Q(p)$, (i.e., for continuous distributions $x = Q(p)$ if $\text{Prob}(X \leqslant x) = F(x) = p$). The inverse probability law says that, if u is distributed uniformly on the internal (0,1), then $Q(u)$ has the distribution F. Thus if we know how to compute Q, we can use pseudorandom fractions to simulate F by the sequence $\{Q(u_i)\}$.

Another simple general method is *acceptance sampling* (or rejection sampling, depending on one's attitude). This method assumes that a scaled version of the density function, $f(x)$, fits entirely underneath a density function $f_+(x)$ for some other distribution, F_+. For the method to be practical f_+ must be easier to simulate than f. The method is to generate a value, say x', from the distribution F_+, and then a uniform u' on the range, $(0, f_+(x'))$. If $u' \leqslant cf(x')$ where $c < 1$ is the scale factor, then the routine returns $x = x'$. Otherwise a new x' is generated and so on. That the method works can be seen by considering

$$\begin{aligned}
&\text{Prob}(x' \leqslant x_0 \text{ and } x = x') \\
&= \int_{-\infty}^{x_0} f_+(x') \int_0^{cf(x')} (f_+(x'))^{-1} du' dx \\
&= c \int_{-\infty}^{x_0} f(x) dx.
\end{aligned}$$

Since c is the marginal probability of acceptance, i.e., $\text{Prob}(x = x')$, then

$$\text{Prob}(x' \leqslant x_0, \text{ given } x = x')$$

$$= \int_{-\infty}^{x_0} f(x)dx$$

as desired. Figure 1 illustrates the method.

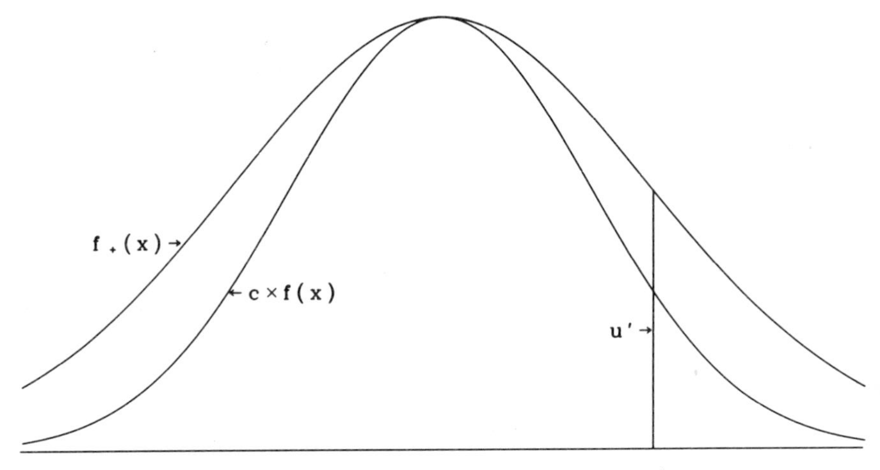

Figure 1: Acceptance Sampling

The cost of each trial is primarily the cost of sampling from F_+ plus the cost of evaluating $f_+(x)$ and $f(x)$. The average cost is the product of the cost of one trial by the expected number of trials: namely, c^{-1}. An efficient acceptance method, therefore, depends on finding an easily simulated distribution that can be scaled to look very much like the desired distribution. For distributions with infinite range, the condition $f_+(x) \geqslant cf(x)$ requires F_+ to have thicker tails than F.

For single distributions neither the inverse method nor acceptance methods are generally as efficient as the method of mixtures described next. However, for some parametric families of distributions, acceptance sampling remains competitive (see Section **g**).

A general technique that does produce efficient algorithms is the *method of mixtures.* Suppose F(x) can be written as the mixture of distributions, $F_1(x),...,F_k(x)$, with F_j being chosen with probability p_j. Then

$$F(x) = p_1F_1(x) + \cdots + p_kF_k(x). \tag{15}$$

The general method is to divide the interval, (0,1), into k subintervals, the jth being from q_{j-1} to q_j, where

$$q_j = \sum_{i=1}^{j} p_i .$$

A pseudorandom fraction is generated and the interval containing it is determined. If this is the jth interval, an observation from $F_j(x)$ is then simulated. The trick behind the use of the method of mixtures is to arrange that F_j can be cheaply simulated whenever p_j is large, so that cheap calculations are done frequently and expensive ones only on rare occasions.

One particular class of such mixtures is generated by approximating the density function, $f(x)$, on successive intervals, by an approximation that always lies between zero and $f(x)$, and that itself corresponds to the density of an easily simulated distribution. The most obvious and simple such approximation is a constant, i.e., the density of a uniform distribution. Provided that $f(x)$ is convex ($d^2f/dx^2 > 0$) the trapezoid formed by joining $(x_i, f(x_i))$ to $(x_{i+1}, f(x_{i+1}))$ also lies below the density. One may then sample directly from the trapezoidal density or, more efficiently, sample from a mixture of the rectangular box and the triangular wedge above it (see Figure 2). At any rate, given intervals $[x_i, x_{i+1}]$ for $i = 0, \cdots ,M$, say, we define approximations $h_i(x)$ to $f(x)$ on the ith interval, with

$$0 < h_i(x) \leqslant f(x) \tag{16}$$

for $x_i < x \leqslant x_{i+1}$. Then if H_i is the distribution with density proportional to $h_i(x)$ on $[x_i,x_{i+1}]$ and 0 elsewhere,

$$F(x) = \sum_{i=0}^{n} p_iH_i(x) + p_{n+1}F_*(x), \tag{17}$$

or writing in terms of densities

$$f(x) = \sum_{i=0}^{n} p_ih_i(x) + p_{n+1}f_*(x).$$

The remainder distribution has density

$$f_*(x) = (f(x) - h_i(x))\times c$$

on the ith interval. Figure 2 shows an approximation of this form. In the tails f_* is proportional to f.

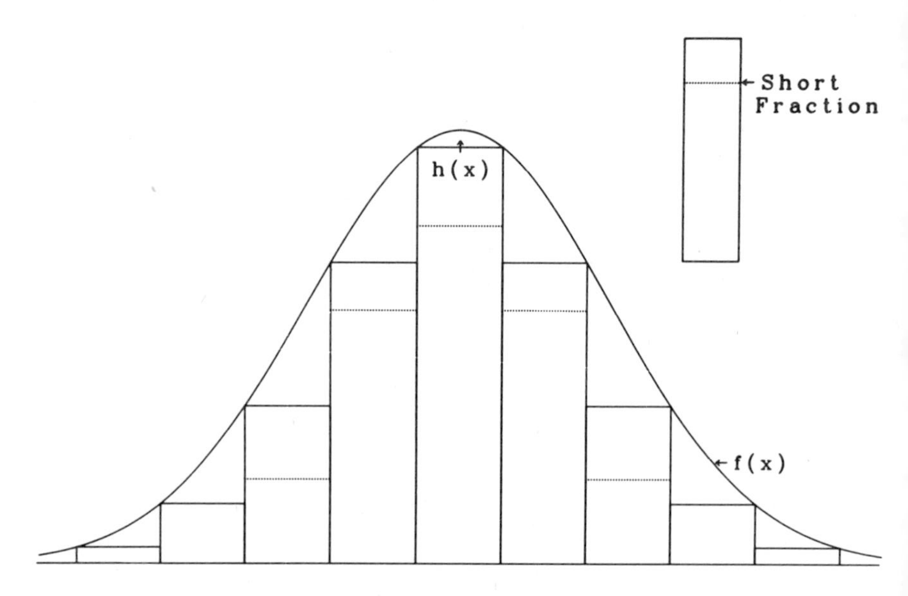

Figure 2: Method of Mixtures

Somewhat more refined techniques are possible. In particular, one can choose the x_i and the p_j to be "short" binary fractions, so that we can select one of the uniform distributions and sample from it using the bits in a pseudorandom integer, with a minimum of arithmetic. The details are beyond our scope here: see Knuth (1969, 105-112) for the case of a normal distribution. This method, originally applied by Maclaren and Marsaglia to the normal distribution, can be used for a general distribution. For example, see Maclaren, Marsaglia, and Bray (1964) for the exponential distribution. The design requires being able to simulate the remainder distribution occasionally, using one of the inefficient general procedures or some special technique for the distribution in question.

Other choices for the approximating functions, $h_i(x)$, might give a better fit to f(x) while still being easy to simulate. If such techniques reduce the probability of sampling from $F_*(x)$ enough, overall efficiency is improved. As the approximating function becomes more complex, the constraint that $h_i(x)$ be between 0 and $f(x)$ is more difficult to satisfy.

Methods have been proposed for *approximately* simulating general distributions. One is just a variant of (17) but ignoring $F_*(x)$ and not requiring (16) to hold. For example, one could approximate $f(x)$ over a finite range by a piecewise linear function, and simulate $H(x)$ from its inverse; namely,

$$Q_H(p) = x_j + (p - q_j)(x_{j+1} - x_j)$$

where $q_j = \sum_{i=0}^{j-1} p_i$ and $q_j < p \leqslant q_{j+1}$. A difficulty with such techniques is that high accuracy may be obtainable, if at all, only with a very large number of intervals.

A classical statistical technique for intractable problems is to approximate the given distribution by a member of a tractable family. Ramberg and Schmeiser (1972) use Tukey's λ-distribution in a four-parameter form, with inverse distribution function

$$Q(p) = \lambda_1 + (p^{\lambda_3} - (1-p)^{\lambda_4})/\lambda_2 \, . \tag{18}$$

Moments of the distribution can be developed in terms of beta functions and used for fitting the λ_j. Other criteria, using density functions, etc., are not very tractable, since one can easily compute the density as a function of p, but not of x.

Such general methods with limited accuracy introduce an added source of error into simulations. The effect of such errors is hard to estimate precisely. For this reason, it seems desirable to use such methods only when precise answers are not an essential feature of the simulation.

A discrete distribution taking on a sufficiently small set of values, say $1, \cdots, N$, such that we can store tables of size N, can always be simulated from a single uniform by a comparison with tables of q_j where

$$q_j = \sum_{i=1}^{j} p_i \, .$$

To speed up the arithmetic, the method of mixtures can be applied, approximating the p_i by short binary fractions. When the set of values is too large to store the probabilities, some algorithmic process must be used, and the situation becomes more like that for continuous distributions.

g. Special Distributions.

While the method of mixtures, in one form or another, promises a reasonably general method of producing efficient simulations, there are some problems in applying it. When the distribution depends upon certain parameters, the approximating distribution, H(x), also depends on these. In most cases, the appropriate choice of H(x) must be redetermined whenever new parameter values appear. This makes the method less attractive if parameter values change frequently.

Dependence upon parameters can sometimes be eliminated. If there is some function $x = x(t,\lambda_1,\lambda_2, \ldots, \lambda_k)$, where t has a distribution not depending on the λ_i, and the set of t for which $x \leqslant x_0$ is a union of intervals, then the distribution of x given $\lambda_1, \ldots, \lambda_k$ can be simulated by simulating t. The standard case is that of location and scale families

$$t = \lambda_2 \times (x - \lambda_1) \tag{19}$$

and the standard among these is the normal distribution. The fastest, fully accurate method for the normal appears to be Marsaglia, Maclaren and Bray (1964), provided it is implemented in a sufficiently low-level language to take full advantage of machine characteristics. Alternative algorithms, reasonably efficient and easier to implement, are given in Kinderman and Ramage (1976) or Ahrens and Dieter (1974), along with some timing comparisons. Similarly, for the exponential distribution a mixture method described in Maclaren, Marsaglia and Bray (1964), is very fast. See also Ahrens and Dieter (1972).

The distributions for which highly efficient general algorithms exist then are chiefly:

(1) uniform

(2) triangular

(3) exponential

(4) normal

Simulation from these distributions may be done by the procedures noted.

There is a partially general technique with many applications; namely, the method of transformation. Suppose x is some random variable and we can show that x has the same distribution as some function

$$t_x(v_1, v_2, \ldots, v_k)$$

where v_i are other random variables, independently distributed by some k distributions. Then if we can efficiently simulate the v_i, evaluation of t_x will simulate x. This procedure is *not* likely to be as efficient as the best method of mixtures possible, but it has the advantage that t_x may depend upon any number of parameters, $\lambda_1, \lambda_2, \ldots$, and the basic procedure still applies.

The derivation of such a transform requires an analytic mathematical argument. One procedure that can be used is to show that x and k−1 other variables have a joint density, $f_*(x, x_2, \ldots, x_k)$, and that

$$f_*(x, x_2, \ldots, x_k) J(x, x_2, \ldots, x_k) = (\prod f_i(v_i))$$

where f_i is the density for v_i and J is the Jacobian of the transform from the v_i to the x_j.

There are many specific examples, of which the following are notable. In this discussion $u_1, u_2, \cdots$, are uniform on (0, 1), $n_1, \cdots$, are standard normal and $e_1, \cdots$, exponential. The normal itself can be written

$$\begin{bmatrix} n_1 \\ n_2 \end{bmatrix} = (-2 \log(u_1))^{1/2} \begin{bmatrix} \cos(2\pi u_2) \\ \sin(2\pi u_2) \end{bmatrix}. \tag{20}$$

Known as the Box-Muller transform, this gives us two normals from two uniforms and some arithmetic. Although it can be speeded up and trigonometric functions avoided, this is not competitive to the mixture technique. It might be attractive for very small machines because of its simplicity, particularly if special functions and the uniform generator were built in.

The exponential also has a noncompetitive transform:

$$e_1 = -\log(u_1)$$

which is, in fact, just the inverse distribution function. The *Cauchy* or *Lorentz* distribution

$$f(x) = 1/(1 + x^2)$$

likewise has a transform derived from the inverse, this time fairly competitive:

$$x = \tan(\pi(u_1 - 0{\cdot}5)). \tag{21}$$

It seems likely that a mixture method might be developed to beat this, but none has appeared at the time of writing. The Cauchy distribution's long tails (i.e., the relative slowness with which 1-F(x) and F(x) approach 0 as $x \to \infty$ and $-\infty$, respectively) make approximation on a finite interval less effective than for the normal. Notice that (21) is the ratio of the two elements on the right in (20). Therefore, an equivalent form for the Cauchy is:

$$x = n_1/n_2$$

The family of *stable distributions* (Feller, 1971, 574-581), which includes both Cauchy and normal, has a general transformation of which (20) and (21) are special cases. The general result was developed by Chambers, Mallows and Stuck (1976) from earlier work of Zolatarev and others. One form of writing the result is

$$x = \left(\frac{\cos((1-\alpha)\phi + \alpha\phi_0)}{e} \right)^{\frac{1-\alpha}{\alpha}} \frac{\sin(\alpha\phi - \alpha\phi_0)}{(\cos\phi)^{1/\alpha}} \tag{22}$$

where α is the shape parameter of the stable distribution, with $0 < \alpha < 2$ and $\alpha \neq 1$, and ϕ_0 is a skewness parameter $|\phi_0| < \frac{\pi}{2}$. The basic random variables are ϕ, uniformly random on $(-\frac{\pi}{2}, \frac{\pi}{2})$ and e, standard exponential. For better, but less conventional, parameterizations, computational details, and limiting cases, see Chambers et al. (1976). There are also location and scale parameters that can be introduced as in (19). This is a particularly interesting example in that the stables have no closed-form density except for $\alpha = 1$, $\phi_0 = 0$ (Cauchy), $\alpha = 2$ (normal), and $\alpha = ½$.

Two discrete distributions for which reasonably efficient special algorithms exist are the geometric and the Poisson distributions. The distribution function for the geometric distribution is

$$\begin{aligned} F(n) &= \sum_{i=1}^{n} (1-p)^{i-1} p \\ &= 1 - (1-p)^n \end{aligned}$$

It immediately follows that Q(p) is the smallest integer as large as

$$\log(u_1)/\log(1-p).$$

For the Poisson distribution, a simple method of simulation is based on the observation that:

$$\text{Prob(Poisson} = \text{n}) = \frac{e^{-\mu}\mu^n}{n!}$$

can be shown to equal the joint probability that

$$\prod_{i=1}^{j} u_i \geqslant e^{-\mu} \quad \text{for } j = 1,\ldots,n-1, \text{ and}$$

$$\prod_{i=1}^{n} u_i < e^{-\mu} .$$

Thus one can generate uniforms until their product is smaller than $e^{-\mu}$. If many values are to be generated for fixed μ, one would wish to develop numerical values for the probabilities, at least for small μ. Snow (1968) gives such a procedure. See also Knuth (1969, 117-118), and Ahrens and Dieter (1974) for other Poisson methods.

Several distributions are representable as sums of random variables from the distributions just described. For example, a chi-square variable with ν degrees of freedom can be formed from $[\nu/2]$ exponentials and $\text{mod}(\nu,2)$ squared normals. A similar result then holds for F distributions. The binomial distribution is defined as the number of times a uniform variable $u \leqslant p$ in n trials. For most such distributions, however, better methods exist. The gamma and beta distributions (and therefore the special cases of chi-square and F) are discussed in Ahrens and Dieter (1974). They describe a clever acceptance method for the gamma distribution which takes advantage of the limiting normality as the shape parameter becomes large. For small shape parameters, they use the summation characterization, and then apply an acceptance technique for the fractional part of the shape parameter. The same reference also discusses methods for binomial and negative binomial distributions.

h. Multivariate Distributions

The simulation problems of interest in this section are *multivariate* random variables $x = (x_1, \cdots, x_k)$, where the k components of x are *dependent.* The completely general multivariate distribution can be simulated by using *conditional* univariate distributions. If $F_j(x_j \mid x_1, \cdots, x_{j-1})$ represents the probability that the jth component of x is less than x_j given the values of $x_1, \ldots, x_{j-1}$, then

$$F(x_1, \ldots, x_k) = \prod_{j=1}^{k} F_j(x_j \mid x_1, \cdots, x_{j-1}).$$

Thus assuming that each of the F_j can be simulated as univariate distributions for x_j, a k-step algorithm for generating x is to first generate x_1 from $F(x_1)$, then given x_1 generate x_2 from $F(x_2 \mid x_1)$, and so on. Needless to say, this is an expensive and difficult process in general.

There are a few special classes of multivariate simulations for which very simple procedures are possible. Fortunately, and certainly not coincidentally, these are the most commonly encountered multivariate distributions. The first is simply the set of *order statistics* for a sample of k from the uniform distribution. The obvious method of simulating this distribution is to generate $u_1, \cdots, u_k$ as independent uniforms and sort them. For moderate values of k (say up to 100 or so) or if we want all or nearly all the order statistics, this is a perfectly reasonable method. See Section **3.d** and the Appendix for sorting algorithms. Note that partial sorting may be used if only some of the order statistics are required. On computers for which integer comparisons are significantly faster than floating-point comparisons, a more efficient method is to generate k pseudorandom integers and sort these, before converting them to the corresponding floating-point fractions.

Notice that, since the inverse distribution function, $Q(p)$, is a monotone function of p, sorting $u_1, \cdots, u_k$ is equivalent to sorting $Q(u_1), \cdots, Q(u_k)$. As a result we can find specific order statistics of a sample from a general distribution, using the order statistics of the uniform, and evaluation of $Q(u_i)$ for the desired order statistics only.

There is a general distribution for uniform order statistics as follows. Let $j_1 < j_2 \cdots < j_s$ be any set of integers on the range 1 to k. Let v be the vector of s corresponding order statistics from a sample of k uniforms; i.e., the first element of v is the j_1 order statistic and so on. Then we have the following result. If $g_1, g_2, \cdots, g_{s+1}$ are independent gamma random variables with parameters

$$\eta_i = j_i - j_{i-1}$$

for $i = 2, \cdots, s$, $\eta_1 = j_1$ and $\eta_{s+1} = n - j_s$, then the joint distribution of the elements of v is the same as that of the variables

$$v_i = \left(\sum_{r=1}^{i} g_r \right) \Big/ \left(\sum_{r=1}^{s+1} g_r \right) . \tag{24}$$

This provides a mechanism for generating s order statistics from s+1 independent gamma variables. The combination of sample size and number of order-statistics at which this method becomes attractive depends upon the efficiency of the gamma simulation. See the Appendix for some algorithms.

In the special case where *all* the order statistics are wanted, the g_r have exponential distributions that can be efficiently simulated. Thus we have two methods for generating all order statistics. The computational costs of either method should be quite reasonable, assuming both are well programmed. Some published comparisons have appeared, but since both methods were inefficiently implemented, the results are best ignored.

Multivariate Normal Distribution. This is by far the most widely used model for multivariate data and can be simulated quite efficiently in all cases, by use of the linear transformation properties obeyed by the distribution. The related family of Wishart distributions can be simulated by similar procedures. In this discussion we use without comment the basic tools of linear algebra, as discussed in Chapter 5.

The distribution of a general multivariate normal in k dimensions is determined by a k-vector, μ, its mean, and a positive semi-definite k by k matrix, Σ, its variance matrix. In particular when $\mu = 0$ and $\Sigma = I$, the identity matrix, the distribution is just that of k independent standard univariate normals. The essential fact in simulating the general case is the following. Let $z = (z_1, \cdots, z_k)$ be k independent standard normals. Let c and A be respectively an arbitrary k-vector and an arbitrary k by k matrix. Then the random variable

$$x = A' \cdot z + c \tag{25}$$

has a k-variate normal distribution with

$$\begin{aligned} \mu &= c \\ \Sigma &= A' \cdot A \end{aligned} \tag{26}$$

Furthermore, any valid μ and Σ can be represented this way. Therefore, we have a fully general technique for simulating normals; namely, generate k univariate normals, and apply (25).

The choice of A can vary according to the purpose at hand. For example, one could start with a Choleski decomposition of Σ and obtain an upper-triangular A, or with an eigenvalue decomposition of Σ and obtain A as $V \cdot D$ with V orthogonal and D diagonal. Similarly orthogonal or singular-value decompositions of a data matrix would generate A consistent with the sample variance structure.

The distribution of the cross-product matrix of a sample of n multivariate normals is the Wishart distribution. It depends upon n, Σ and $\Delta = M' \cdot M$ where M is the n by k matrix whose i^{th} row is the mean of the ith member in the sample. When $M = 0$, the central Wishart can be simulated by k chi-square variables and $\frac{1}{2}k(k-1)$ unit normals. An additional k^2 normals are sufficient to handle the noncentral case. See Chambers (1970, pp. 9-10) for the specific procedures.

i. Monte-Carlo Methods

The Monte-Carlo method is another name for the process of approximating expected values (i.e., integrals with respect to a probability distribution) by sample means. Thus it is an attempt to do computations by the law of large numbers. Abstractly stated, the method is as follows. Let F be a distribution function on some probability space, Ω, and let g be any real-valued function on the same space. Then the expected value of g with respect to F is

$$\mu(g,F) = \int_{\Omega} g(x)\, dF(x). \tag{27}$$

Then the strong law of large numbers states that, if the expectation of g exists, then with probability 1 the mean of g for a sample of M from F, namely,

$$m(g,F,M) = \left[\sum_{i=1}^{M} g(z_i)\right] / M \tag{28}$$

tends to $\mu(g,F)$ as $M \to \infty$, where z_i are independently distributed according to F.

As a computational technique, the Monte Carlo method is simply the attempt to compute (27) approximately, as the mean of $\{g(z_i)\}$, where $\{z_i\}$ is a sequence of pseudorandom numbers generated from the distribution F. In principle the method applies to any distribution that can be simulated, discrete or continuous. The most common applications are to the case that F is a continuous distribution in some finite number of dimensions, so that

$$\mu = \int_{-\infty}^{\infty} \cdots \int_{-\infty}^{\infty} g(x)p(x)\, dx_1 \cdots dx_k \tag{29}$$

with $x = (x_1, \cdots, x_k)$ a k-dimensional real vector, and

$p(x) = p(x_1, \cdots, x_k)$ the probability density for F. The most specialized form of the problem is that of uniform rectangular estimation:

$$\mu = \int_0^1 \cdots \int_0^1 g(x)\, dx_1 \cdots dx_k \tag{30}$$

where the components of x are all identical independent uniform variables on (0,1).

On the other hand what we have called the Monte Carlo method is just a special case of the simulation of a random process. If $\{z_i\}$ is a sequence of random variables, not necessarily independent, if $F^{(M)}$ is the distribution of $z_1, \cdots, z_M$, and if the sequence, $\{g^{(M)}(z_1, \cdots, z_M)\}$, converges with probability one to $\mu = \mu(\{g^{(M)}\}, \{F^{(M)}\})$ then we can estimate μ by generating a pseudorandom sequence, $\{z_i\}$. This type of extended simulation arises, for example, in the study of time-series or stochastic-process models, where μ is some long-term trend or limit of the process. Computationally, there is rarely much new in the extended form as we tend to simulate dependent processes by some form of independent increments. There is, also, no need to restrict g to be real-valued; it could be vector-valued, for example, with trivial change in the computations.

For the most part, Monte Carlo calculations are the approximation of integrals. Relative to ordinary numerical integration techniques, three main advantages may exist.

(i) As the dimensionality of the integral increases, the cost of Monte Carlo estimation increases much more slowly then that of iterated univariate integration.

(ii) Particularly efficient choices of distribution and/or specialized variance reduction techniques may reduce costs.

(iii) For generalized simulation problems, there may be no obvious integration alternative.

The Monte Carlo procedure tends to be particularly suited to multidimensional integration and to the simulation of complex systems.

When an integral is computed by Monte Carlo, one will obviously not have the same type of error bounds as for conventional integration techniques. Instead one may use the distribution simulated to support probabilistic statements about the error. The mean of M simulations in (28) has the distribution of the mean of M random variables with the distribution of g(z). Assuming this distribution to have a finite variance, the central limit theorem applies:

$$\frac{m(g,F,M) - \mu}{\sigma/\sqrt{M}} \xrightarrow{L} N(0,1) \tag{31}$$

so that the estimate, $m(g,F,M)$, has a limiting normal distribution, mean μ and variance σ^2/M, where

$$\sigma^2 = \int (g(z) - \mu)^2 dF(z). \tag{32}$$

Since σ^2 is unknown, a sample estimate, such as

$$s^2 = \sum_i (g(z_i) - m(g,F,M))^2/(M-1) \tag{33}$$

can be computed and used to estimate the error in $m(g,F,M)$. The standard t-statistic, based upon the M pseudorandom values of z, gives an approximate confidence region (pseudoconfidence region?) of the form

$$m(g,F,M) - M^{-1/2} s\, t_\alpha < \mu < m(g,F,M) + M^{-1/2} s\, t_\alpha \tag{34}$$

where t_α is the value exceeded with probability α by the absolute value of a t distribution with M degrees of freedom. For M typically large the latter is essentially the normal distribution.

Clearly, the smaller the variance, the more accurate the answer, in the probabilistic sense. There exist a large number of tricks designed to reduce the variance. The following are among the more generally useful. The *controlled-variate* technique looks for a function, g_*, such that the expectation of g_* is easy to compute (ideally does not require simulation), and that g_* looks like g; i.e., $g - g_*$ has low variance. Then one estimates μ as

$$\int (g-g_*)\, dF + \int g_*\, dF.$$

The *importance sampling* technique works on F the same way. Suppose F has a density $p(z)$, and we can find a density p_* that is not difficult to sample, and such that $g(z)p(z)/p_*(z)$ has low variance. One then estimates μ as

$$\int \{g(z)p(z)/p_*(z)\}p_*(z)\, dz.$$

One can try to modify the range of integration along the same lines. Here there are several techniques. The *antithetic variate* technique, in its simplest form, breaks the range, Ω, into two exclusive subsets, Ω_1 and Ω_2, such that F is symmetric between them, in the sense that

$$\int_{\Omega_2} g\, dF = \int_{\Omega_1} (-g_*)\, dF$$

for some g_* derivable easily from g. If $g - g_*$ is of low variance, then μ is estimated by

$$\int_{\Omega_1} (g - g_*)\, dF$$

The method extends to more than two subsets. The most obvious and probably most useful case is that of integrating (27) along a dimension in which g is monotone, so that it is profitable to replace, for example, $\int_0^1 g(z_1, \cdots)\, dz_1$ by:

$$\int_0^{1/2} (g(z_1, \cdots) - g(1 - z_1, \cdots))\, dz_1.$$

A class of methods that may be termed *conditional sampling* relies on observing that, if we fix the value of some variable, say z_*, we can easily compute the conditional expectation

$$\mu_*(z_*) = \int_{\Omega_*} g(z)\, dF_c(z|z_*)$$

where F_c is the conditional distribution and Ω_* the range of integration with z_* fixed. Assuming this done, then μ is estimated as

$$\int \mu_*(z_*)\, dF_*(z_*),$$

with F_* the marginal distribution of z_*.

Particularly desirable cases are those in which the conditional expectation can be derived analytically or in which the different values of μ_* are related in a simple way, such as by translation and/or scaling. The specific case that z is a real variable defined as

$$z = x/y$$

with x standard normal and y some positive variate has been very successful (Relles, 1970; Gross, 1973).

These and numerous other techniques are clearly part of the art of computing. The wise choice of variance reduction techniques must come from knowledge of the problem, experimentation, and study of similar problems. The interested reader is referred to

Halton (1970) or Hammersley and Handscomb (1964) for some basic discussions. Hastings (1970) extends several of the techniques to more general stochastic processes.

j. Summary and Recommendations

Pseudorandom values are usually generated to approximate expected values (integrals with respect to some probability measure), using averages from repeated simulation (Monte Carlo). In problems of some complexity, the validity of the approximation cannot be established from easily verifiable properties such as equidistribution. Further, the more general goal of replicating entirely the properties of a random distribution is not attainable algorithmically.

For the simulation of the uniform distribution, a reasonable balance between good statistical properties and computational simplicity is achieved by one of the methods of postprocessing the output of simple generators.

In addition to the general techniques for other distributions discussed in Section **f**, some special techniques have been described. With the exception of the few explicit programs noted, the references contain verbal or mathematical descriptions of algorithms. The methods are, as far as current evidence goes, competitive in speed and accuracy for the distributions cited, but at the time of writing there are few thorough, accurate, and unbiased studies of sufficient generality to apply across differing machines and computing environments. See the Appendix for references to algorithms for various distributions.

Finally, some questions of portability and reproducibility need to be considered, applying to all the topics discussed in this chapter. The results of simulations function as the "experimental reports" of mathematics and statistics. Therefore, the requirements of verification and reproducibility applied to experimental science are equally relevant. Unfortunately, practical problems and carelessness have often cast doubt on the reproducibility of simulation results.

The nature of the algorithms described in Sections **b** to **e** make their efficient implementation in FORTRAN difficult or impossible. More basic is the fact that the definition of the generator itself usually takes the word length, and possibly other machine characteristics, as given. The following general approach avoids this problem, and should be considered carefully. Given a specific generator on a particular machine, the integer sequence produced can be emulated exactly on any other machine as follows.

Consider the sequence of generated integers, r_i. Then each r_i can be represented as a vector of k short integers, $s_{i1}, \ldots, s_{ik}$:

$$r_i = \sum_j s_{ij} \beta^{j-1}$$

where β is the base of the short integers. By emulating the calculations of Sections **c** to **e** on the vectors, the exact sequence, r_i, is obtained. Further, if β is short enough; e.g., $\beta = 2^{16}$, the computations can be done by a portable FORTRAN algorithm that runs on nearly any computer. (One may eventually choose to code a faster assembly language version. This has nothing to do with the emulation issue.)

Fractions can then be computed from

$$u_i = \sum s_{ij} \beta^{j-1-k} .$$

The precise value of u_i will depend on the machine. Therefore it is essential that any derived distribution using an acceptance rule or other choosing mechanism work with the integers, s_{ij}, directly. Given these cautions one may produce essentially equivalent simulations on any computer. See Gross (1976), the Appendix, and Problem 2 for a FORTRAN emulator of the McGill generator and for further discussion.

Problems

1. One idea behind the shuffled generators described in Section **e** is that permuting a block of M numbers may break up subsequences of length, k, where $k << M$. Derive a numerical estimate of how well this is done, on the average, by a permutation. Hint: use the distances among the first k elements after permuting, the next k, etc.

2. (Gross, 1976) Construct a high-level-language implementation of the McGill generator.

   (a) Implement a 32-bit congruential generator $\{r_i\}$, in a high-level language, using the relation

   $$r_{i+1} = \text{mod}(r_i \times 69069, 2^{32})$$

   with r_i stored in a vector of integers as suggested in Section **j**. Hint: $69069 = (13 \times 64^2) + (55 \times 64) + 16$.

(b) Implement a 32-bit shift register generator, $\{r_i'\}$, defined by

$$r_*' = \mathrm{XOR}(r_i', \mathrm{SHIFT}(r_i', -15))$$

$$r_{i+1}' = \mathrm{XOR}(r_*', \mathrm{SHIFT}(r_*', 17))$$

where the bit operation, SHIFT, shifts the first argument (left or right as the second argument is >0 or <0), and the operation, XOR, performs an exclusive-or operation of the two arguments. (See Section **4.a** and Problem 4.1) In this application, it is not neccesary to conserve space, and the easiest implementation is to represent the bits by a logical (boolean) vector of length 32.

(c) Compute the result

$$r_i'' = \mathrm{XOR}(r_i, r_i')$$

and the fraction, $u_i = r_i''/2^{32}$. Consider how to do this from the short integers, in case 2^{32} cannot be represented in the machine.

3. Design an algorithm to simulate a random point on the surface of a sphere in k dimensions, for arbitrary k. Such a point can be described by k−1 angles, θ_j, where θ_j is the angle from the jth co-ordinate axis to the point on the sphere. The point is uniformly distributed if the probability of a surface patch is proportional to its area. In this case, the angles are independent and uniformly distributed. Show that the transformation to Euclidean co-ordinates is

$$x_j = \cos\theta_j \prod_{i<j} \sin\theta_i, \quad j=1, \cdots, k-1$$

$$x_k = \prod_{i=1}^{k} \sin\theta_i$$

A fairly efficient, easy method comes from generating k values, $n_1, \cdots, n_k$, from the standard normal distribution. Show by a transformation argument that the desired values are given by $x_j = n_j / r$, where $r^2 = \sum n_j^2$. This is trivial to program and not very expensive. An elegant, non-obvious method due to Marsaglia (1972a) is claimed to be more efficient in three and four dimensions, but a general algorithm for arbitrary k is not straightforward.

4. Design a "meta-algorithm" to simulate N observations from an arbitrary family of distributions. Your algorithm will be given N and a function, Q, to compute quantile values for any member of the family. Suppose that an algorithm is available, also, that will generate an approximation to Q as a function of P for a *fixed* member of the family (cf. Section **4.e**). Your algorithm now has the choice of applying Equation (14) directly, or with Q replaced by a more efficient approximation. What estimates of cost would be needed to choose, and how would they be used?

CHAPTER EIGHT

Computational Graphics

a. Graphics for Data Analysis

This chapter has two purposes: to present some important general techniques in computer graphics and to discuss the special requirements of data analysis. This introductory section presents an overview of graphics for data analysis. Sections **b** to **g** discuss general techniques, and the remaining sections suggest how to apply them to data analysis. Readers concerned with basic graphical hardware and software are directed to the earlier sections. Those interested only in building procedures for data analysis on existing basic procedures may skip to the relevant later sections.

A number of graphical software systems are available. Most large manufacturers of terminals provide program packages, and some commercial software vendors also market packages. If such packages are to provide powerful, flexible facilities, they must make the common graphical operations in data analysis easy to perform. Such operations include scatter plots, histograms, probability plots, and other more specialized plots. In addition, the systems should allow users to modify or extend the standard operations and apply them in a variety of situations.

The concepts outlined in this section and referred to in the rest of the chapter are based on a system of graphical subroutines developed at Bell Laboratories (Bell Laboratories, 1977; Chambers, 1975; Becker and Chambers, 1976). This is a portable, structured set of FORTRAN subroutines, specifically designed for data analysis. At the same time, we feel that the concepts must, in some form, be part of any reasonably advanced graphical system to be used in data analysis.

Graphical output is described as a series of *figures.* Each figure consists of a central *plot* in which data, curves, or other statistical information are presented. In the four *margins* surrounding the plot, additional information, such as titles or axis labels, may be provided. Figure 1 illustrates the concept. The generated figures occupy all or

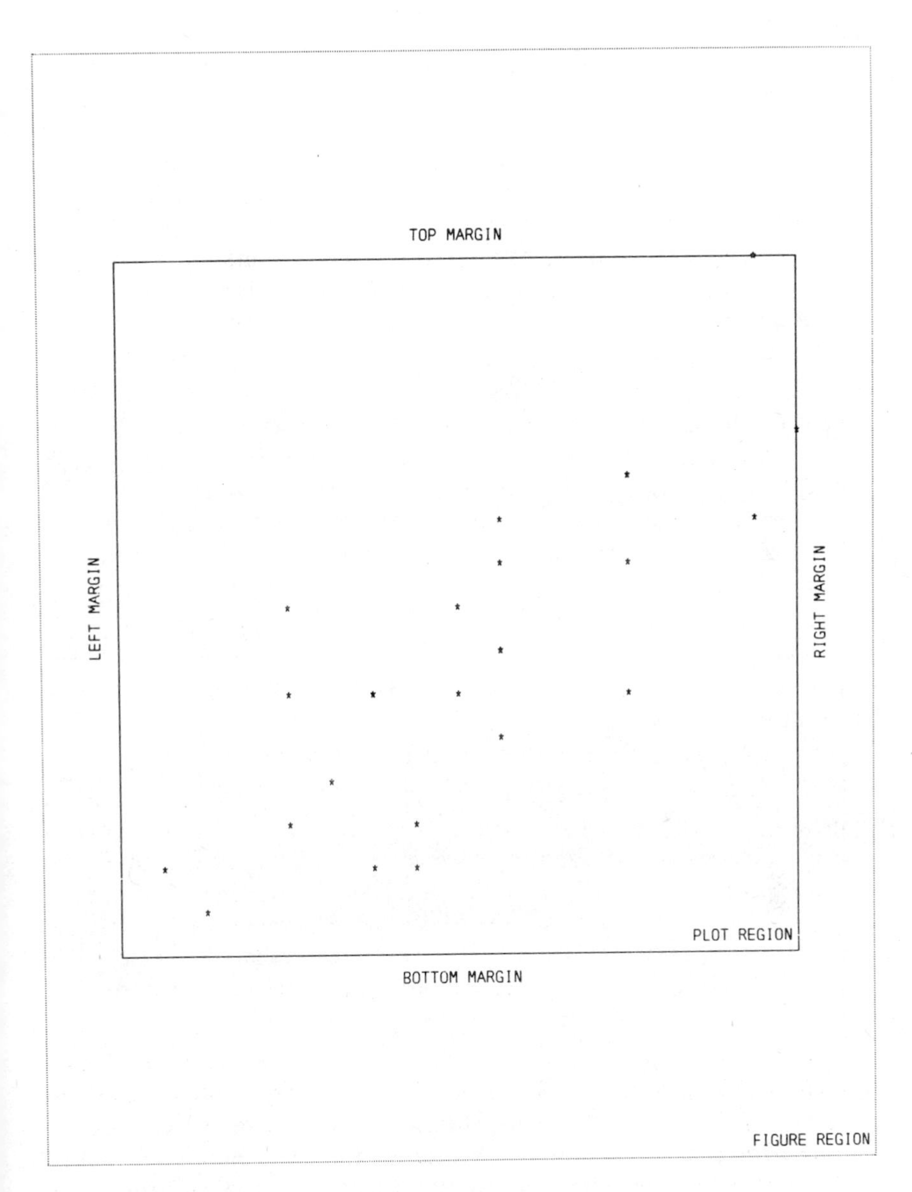

Figure 1: Graphical Structure for Data Analysis

part of a page of output (or of the surface of a terminal device). For example, figures may be arranged as a one- or two-way array on the page.

The concepts of figure, plot and margins help specify where on the page user-defined information is to appear, and provide a natural coordinate system for such information. For data to appear in the plot, the fraction of the page defined by figure and plot forms a viewing window (see Section **c**). The natural user coordinate system for the plot will be based on the range of user data values. Optionally, this range may be adjusted to provide pleasing intervals for axis labels and/or to ensure that data falls strictly inside the plot. (see Section **h**).

For producing text in the margins, a somewhat different coordinate system is suitable. A position in the margins is identified by side (bottom, left, top or right), by line (the distance from the inner edge of the margin, measured in lines of text) and by the position along the side (either in terms of user coordinates or as a fraction of the length of the side). Such a coordinate system simplifies the labelling of axes, generation of titles, or plotting of other auxiliary information.

The mapping from the range of user or margin coordinates (called the data region in Section **c**) to the ultimate device coordinates is a linear transformation, derived in Equation (5). The details of the eventual transformation will generally be of no concern to the user of a higher-level graphical system.

To have flexible control over graphical output, the algorithms need a way to define the detailed interpretation of the various concepts discussed; for example, the size or shape of the plot, the number of lines of margin or size and style of lettering. Such quantities may be made into *graphical parameters,* whose values may be queried or specified by users' programs. Reasonable default choices will be established by the graphical system. Special applications may then override the parameters of interest to them.

Given such an organization, it is relatively easy to design high-level algorithms for scatter plots, histograms, or other statistical plots. These may be constructed from simple sequences of intermediate-level graphical operations for the choice of scales, titling, labeling, and the like. By the use of the graphical parameters and the organization described above, it is possible to write both high-level and intermediate-level algorithms in a clean, modular way. The writer of graphical software is assisted in writing understandable, well-designed programs (compare Section **2.c**). Applications of graphics to research in statistics and data analysis benefit particularly from the structuring of high-level algorithms, since new graphical techniques may be implemented rapidly and with pleasing results from existing intermediate-level operations.

The system design outlined here is only one of many possible, and differs in important respects from comparably sophisticated software for applications other than data analysis. Experience in statistics and data analysis, using the system written at Bell Laboratories based on these concepts, has convinced us that well-designed, portable, modular graphics software can significantly increase the power of graphical methods of analysis.

b. Graphical Devices and Their Capabilities

From the viewpoint of the designer of graphical terminals there are many engineering properties which must be considered and balanced in developing a terminal for a given application. As far as the user or purchaser is concerned, however, specific questions of materials, construction, etc., are important only in their effect on the operational usefulness of the terminal. For this purpose one can list categories of device performance. A comparison of alternative devices in these categories allows one to assess relative merits for a specific job. We shall consider the following four categories:

1. graphical quality,
2. dynamic response,
3. permanence,
4. sophistication.

The actual hardware used currently comprises two basic media: paper, film, plastic or other *material* on which the graphic output is printed or drawn; and *display terminals,* most frequently using cathode-ray tube techniques. Broadly, there is a parallel distinction between the mode of use, in that terminals are more often interactive than are material devices.

Figure 2 describes some of the types of graphical devices available at the time of writing, along with their basic operating principles. Generally, all the displays of types 4, 5 and 6 use the same underlying principle: phosphors coat the inner surface of a display screen and are excited to glow by a movable beam of electrons. One may imagine the beam coordinates as continuous, at least at the

	Device Type	Display Technique	Examples
Material	1. Character	Graphical information coded by characters on the page	Printer, teletype
	2. Pen—Plotter	Marker drawing continuous lines on the material	CALCOMP, GERBER
	3. Photographic	Copied from a display or written directly on film	XEROX—STARE, FR—80
Display	4. Variable Scan	Moving beam continually refreshing the displayed points	Oscilloscopes
	5. Storage Scope	As 4, but once excited, each point remains illuminated indefinitely	TEKTRONIX
	6. Fixed scan or addressable	Each point on the screen is scanned at a fixed frequency, or is directly addressable	Television; Plasma panel.

Figure 2: Graphic Devices

level of hardware design. For programming purposes it is more common, at a low level, to imagine the display as a rectangular array of *rasters,* discrete points which may be excited. For some displays, such as television monitors, the *intensity* of excitation may be varied; for others, it is simply on or off. Color displays of types 4, 5 or 6, generally have three sets of phosphors for the three primary colors (red, blue and green). The intensity of a point is then defined by a triplet.

For the remainder of this section, we examine the use of the various devices in terms of the four criteria above. Remember that rapid advances are taking place in the design of terminals, and the prospective purchaser should obtain as much technical advice as possible.

Graphical quality. The visual quality of a display may depend upon a combination of three related properties: resolution, complexity and accuracy. Resolution is the smallest possible distance between the centers of two adjacent graphic marks. In a *raster* display, we have an array (usually rectangular) of displayable dots. The size of each dot represents a limit on the visual resolution (the precise *effective* resolution is a bit subtle; see, e.g., Knowlton and Harmon (1972)). In a pen-drawn display we have a moving marker, which may actually be a pen moving over paper, an ink-spraying, photographic, light-beam, laser or other movable beam.

Resolution is the most obvious but not the only aspect of graphical quality. The *accuracy* of the plotting determines, for example, whether lines can be drawn to a specified physical length on the device. High accuracy is important when the output is to be used, for example, in diagrams of optical or electronic devices. For data analysis we might be willing to sacrifice such accuracy in most cases, provided that the relative lengths in the plot were not seriously distorted. Photographic recording in type 3 is particularly subject to small inaccuracies.

The quality of plots of type 4 is limited as regards the *complexity* of the displayed picture. When the time to scan the full picture becomes significantly long compared to the *persistence* (defined, say, as the time in which brightness drops to half), the picture will show annoying flicker. Typical persistences are of the order of 3×10^{-2} seconds; much longer would limit the dynamic response of the tube.

Dynamic Response. Two questions need to be asked. How fast can a picture be plotted? How fast can it be changed? The questions are directly relevant only for interactive displays. The mechanical plotters tend to be slowest in plotting and cannot be changed in a direct way. The physical plotting rate in types 4 to 6 is determined by the speed with which the beam can be deflected. By and large, this speed is fast enough that the picture appears instantaneously. The limitation in speed is usually in transmitting the *definition* of the picture to the terminal. The distinction is clear if we consider one implementation of a type 6 display. A television monitor is scanned at standard rate (30 times a second in the U.S.A.) with the intensity of each raster encoded by one or more bits in a digital memory. This process is technically feasible and leads to a system capable of apparently instantaneous display. However, if the displayed picture is the result of fairly extensive calculations, it may not be possible to transmit the definition to the digital memory fast enough to give this effect.

The same argument applies to changing the picture, but here the types must be distinguished. Storage scopes cannot be selectively erased, and the flash that accompanies complete erasure effectively prohibits continuous change of the display. Variable-scan displays are at the opposite extreme: there is essentially no difference between plotting and changing, provided we can redefine the picture quickly enough. In particular, it is straightforward to flash between different pictures or subpictures. Fixed-scan displays are in between. Generally each *change* in the picture must be transmitted. If the number of these is fairly small we can get good dynamic response, even for very complex pictures. Otherwise, the changes will be slow.

Permanence. Opinions differ among users of interactive displays as to the need for a permanent-copy facility. The author finds this essential, but the type of application, other system facilities and the programming style will affect one's attitude. The use of a terminal of types 1 to 3 may not be desirable, for some applications. In this case, permanent copy may be obtained either by a hardware device, possibly shared by a numeber of terminals, or by generating a separate picture with similar software on a device of type 1, 2 or 3. The first alternative is the most convenient. If the cost of a built-in copier is not prohibitive, the user may generate copy easily and fairly reliably. Use of an ordinary camera introduces the usual possibilities of error in exposure, focus, etc., as well as being less convenient. It is, of course, often cheaper, particularly if the camera equipment is available for other reasons.

The generation of a parallel plot off-line, sometimes called *deferred plotting,* is useful when high-quality or high-volume permanent copy is needed. The simplest procedure is to intercept all the graphical calls to plot text, lines, or any other graphical output. The arguments to the plotting routines, plus a code identifying which routine was called, are stored as data. A post-processor later reads this data and calls the corresponding actual plotting routines. The routines may be intercepted at a very low level, to make post-processing simple. Greater flexibility and better tuning of the deferred plot result, however, from intercepting calls in terms of the high-level coordinates mentioned in Section **a**. One needs to intercept parameter changes as well, in this case. Then the post-processor can interpret the graphic calls appropriately for the off-line device.

Sophistication. Much current development centers on so-called *intelligent* graphical terminals. These have significant local computing ability, in terms of a micro-computer built into the terminal, say. When used with a device of type 4, for example, the terminal itself can keep the display refreshed from a locally stored definition. Much more complex pictures can be kept without flicker than a single connection to a remote computer would normally allow. As the terminal becomes more sophisticated other features may be taken over, such as variable projection or rotation of multidimensional plots or interactive execution of graphical commands from the user.

Intelligent terminals seem certain to increase in importance. At the time of writing, their impact on data analysis has been significant but not wide-spread, largely because their cost has been several times greater than for simple terminals (frequently an order of magnitude greater). When one or a few terminals can serve a reasonably large group, or when the cost of the terminals is not considered significant, a sophisticated terminal is recommended, particularly for interactive exploratory data analysis. When many terminals are needed and budgetary constraints are active, data analysis has, perhaps, less pressing need for sophisticated terminals than some other applications.

c. Geometry of Plotting: Two Dimensions

There is a standard set of problems related to the "where" of plotting: the determination of the positions on the display surface at which points, text, lines, curves and other plotting objects are to appear. The simpler of these, which must be solved by any general graphical software, concern the transformation of coordinates from those used to describe data to the internal coordinates that can be fed directly to the plotting device. Higher-level operations may also arise, such as the generation of smooth curves, the geometrical transformation of objects, and the display of surfaces by various methods.

The simplest problem is to translate from the user's coordinate system. For this purpose one usually assumes that a rectangular region is defined such that all the data to be plotted have values on that region; e.g.,

$$D = \{(x,y):\ d_{11} \leqslant x \leqslant d_{12};\ d_{21} \leqslant y \leqslant d_{22}\}. \qquad (1)$$

Let us call D the *data region.* We then propose to map D into a rectangular portion of the display surface; say

$$W = \{(u,v):\ w_{11} \leqslant u \leqslant w_{12};\ w_{21} \leqslant v \leqslant w_{22}\} \tag{2}$$

where (u,v) is a point on the display. We call W the *viewing window.* If graphical programs are to be written in a modular form, so that changing from one display device to another is easy, the viewing window should be treated at some stage in a device-independent manner. Thus, for example, the user could treat the entire display as the unit rectangle $0 \leqslant u \leqslant 1$, $0 \leqslant v \leqslant 1$, and specify W as a subrectangle of this region. Internally, one must eventually convert to device-dependent numbers, which we call *raster coordinates* (even though they may refer to a pen-driven plotter, for example, rather than physical raster dots). Suppose the raster coordinates for the *complete* display are the rectangle

$$R:\ \{(s,t):\ r_{11} \leqslant s \leqslant r_{12};\ r_{21} \leqslant t \leqslant r_{22}\}. \tag{3}$$

The usual problem is to translate user's coordinates into raster coordinates for plotting. Given the regions defined by D, W, and R the necessary linear transformation can be described, for example, by the equations

$$\begin{aligned} h &= t_{11} + t_{12}x \\ v &= t_{21} + t_{22}x \end{aligned} \tag{4}$$

where (h,v) are the raster coordinates corresponding to (x,y). The linear transformation is defined from (1), (2) and (3) to be:

$$\begin{aligned} t_{j2} &= (r_{j2} - r_{j1})(w_{j2} - w_{j1})/(d_{j2} - d_{j1}) \\ t_{j1} &= r_{j1} + w_{j1}(r_{j2} - r_{j1}) - t_{j2}d_{j1}. \end{aligned} \tag{5}$$

This formula results from defining t_{j2} as a scale factor to change lengths in the data window to raster lengths. Given t_{j2}, t_{j1} is defined by requiring the lower left corner of D, i.e., (d_{11},d_{21}), to map into the lower left corner of the viewing window in raster coordinates.

Example: Suppose we have a plotting device whose raster coordinates go from 0 through 2047 in both directions. We wish to draw a plot whose horizontal and vertical data values go from 1950 through 1975 and from -3.14159 through +3.14159, respectively. The plot is to fill the upper left quarter of the screen. Then,

$$D = \begin{bmatrix} d_{11} & d_{12} \\ d_{21} & d_{22} \end{bmatrix} = \begin{bmatrix} 1950. & 1975. \\ -3.14159 & 3.14159 \end{bmatrix},$$

$$W = \begin{bmatrix} 0. & 0.5 \\ 0.5 & 1.0 \end{bmatrix}$$

$$R = \begin{bmatrix} 0. & 2048. \\ 0. & 2048. \end{bmatrix}$$

The appropriate transformation of coordinates is then

$$T = \begin{bmatrix} t_{11} & t_{12} \\ t_{21} & t_{22} \end{bmatrix} = \begin{bmatrix} -79872. & 40.96 \\ 1536. & 162.97 \end{bmatrix}$$

Then, for example, the data point (1960, 1.5708) is mapped into the raster point (h,v) with

$$h = -79872. + 40.96 \times 1960. = 409.6$$

$$v = 1536. + 162.97 \times 1.5708 = 1792.0$$

One further set of values may enter the user's calculations: the physical size of the display surface. When the display device is a pen-type plotter, one may wish to generate a plot of a given actual size; e.g., 22 cm by 28 cm. For a number of such plotters, in fact, "raster" units are actually lengths in some scale. For *any* plotting device one may be interested in the *aspect ratio;* i.e., the ratio of actual lengths in the vertical to horizontal directions corresponding to unit raster lengths. This information is required, for example, if circles are to plot as circles rather than as ellipses and squares as squares rather than as rectangles. For either requirement, it is sufficient to know actual lengths, a_1 and a_2, of 1 raster unit in the horizontal and vertical directions. For example, one could generate a square viewing window by requiring W to satisfy

$$\frac{(w_{12} - w_{11})(r_{12} - r_{11})}{(w_{22} - w_{21})(r_{22} - r_{21})} = \frac{a_2}{a_1} .$$

The discussion so far has used the ordinary Euclidean system of coordinates. Since most displays are rectangular in shape and addressed by horizontal and vertical position, this is in a sense the natural coordinate system. Other systems arise at times either because data is naturally expressed in them or because they have computational interest. *Polar coordinates* are fairly common, with a data point, P, defined by (r,θ), where r is the length and θ the angle, counter-clockwise from the horizontal, of the line joining P to a fixed center, C. If C is the point (c_1,c_2) in Euclidean coordinates, then P is the point (x,y) with

$$x = c_1 + r \times \cos(\theta)$$

$$y = c_2 + r \times \sin(\theta)$$

and we can then substitute in (4) to get raster coordinates.

Homogeneous coordinates are sometimes recommended for their mathematical elegance and for some computational advantages. In this scheme, a point on the plane is represented as any of the triplets (hx, hy, h) for $h \neq 0$. The advantage of homogeneous coordinates is that a 3 by 3 matrix multiplication then can represent a linear transformation in Euclidean terms *plus* a translation and/or projection down onto a line. We consider the essentially analogous situation in three dimensions in the next section, since the projection operation is much more important in that case.

d. Geometry of Plotting: Three or More Dimensions

The overwhelming majority of display devices are planar surfaces, or at any rate we plot on them as if they were. As a result, when data is naturally represented in terms of three- or higher-dimensional coordinates, the central computational problem is to transform down to a planar surface in some natural way. This problem has been attacked for centuries by engineers, architects and artists. As in many areas, the power of the computer frees us from many constraints and suggests insightful new techniques.

Although there is little difference between three dimensions and higher dimensions computationally, the availability of our intuition makes the former easier to explain. The simplest general technique is that of projection through a viewing window. It may be described roughly as follows. An observer stands with his eyes at O. He regards a portion of three-dimensional space through a window; mathematically, a rectangular region, W. Any point, P, visible at O defines a line, OP, intersecting W at P', as in Figure 3.

Based on the concept of light rays passing from the point P to the observer, one could replace P by a point of appropriate color, size and intensity at P' without visible change to the observer. Thus a representation or *image* of a set of points can be derived by projecting them onto the viewing surface, W, in this manner. This process, usually termed *perspective projection,* was developed in the fifteenth century by Renaissance artists and scientists, notably

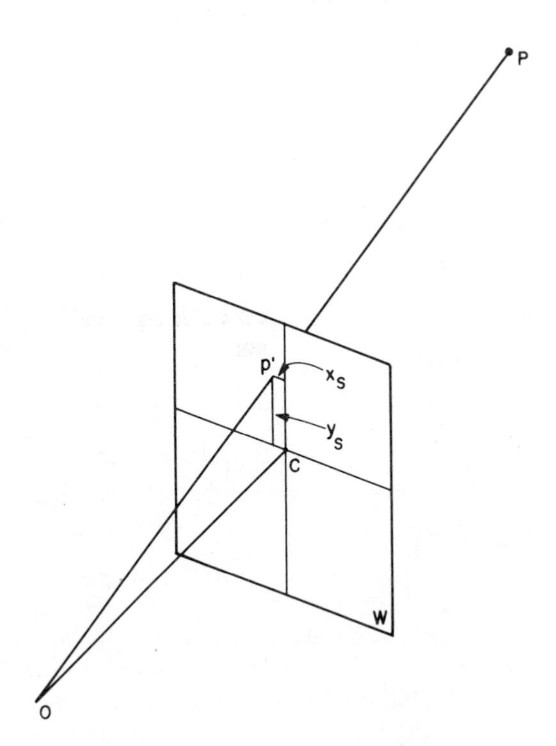

Figure 3: Perspective Projection

Leonardo da Vinci and Fra Luca Pacioli. The rules developed apply in the same way whether W is an artist's canvas or a computer display, except that in the latter case it is convenient to consider the equivalent numerical transformation.

There are a number of formulations of the geometry involved. Most of the published discussions are oriented towards the display of objects and towards the manipulation of such objects, as would be natural in computer-aided design or commercial graphics. For purposes of data analysis, one may wish to use a slightly different formulation. We assume that the observer is at point O, lying on the perpendicular from the center of the viewing window, W.

The point P projects into P'. In order to plot P' we need its coordinates in relation to the center C of W; i.e., (x_s, y_s) in Figure 3. The plane of W is the set of points perpendicular to the line from O to C; i.e., it has equation

$$(X - C)\cdot(O - C) = 0. \tag{6}$$

Since all the points on the line from O to P are of the form, $\alpha P + (1-\alpha)O$, with $0 \leqslant \alpha \leqslant 1$, the point P' has

$$\alpha = \frac{(C - O)\cdot(P - C)}{(P - O)\cdot(P - C)} . \tag{7}$$

(Remember that O, P and C are vectors in three dimensions.) We are free to orient the screen coordinates in any manner within W; i.e., we can take any two perpendicular directions in three dimensions that are also perpendicular to the line OC. In turn, this amounts to a choice of an orthogonal matrix, Q, such that one column is proportional to O–C, say

$$q_3 = Q[,3] = (O - C)/||O - C||.$$

If we let q_1 and q_2, the other two orthogonal directions, be the x and y axes in the viewing window, then the point P' is given by

$$\begin{aligned} P' - C &= x_s q_1 + y_s q_2 \\ &= \alpha P + (1-\alpha)O - C. \end{aligned}$$

Taking inner products with q_1 and then with q_2 gives

$$\begin{aligned} x_s &= \alpha(P - C)\cdot q_2 \\ y_s &= \alpha(P - C)\cdot q_3. \end{aligned} \tag{8}$$

Various special choices of W simplify the calculations. For example, choosing OC to be the z-axis (translated) and leaving the x- and y-axes unchanged gives

$$x_s = \alpha p_1$$

$$y_s = \alpha p_2$$

with $P = (p_1,p_2,p_3)$.

A particular characteristic of data analysis is that the case of three-dimensional points is only one possible situation. More generally, we want to view some projection of points, P, in k dimensions onto a plane. Whereas in three dimensions we could define W by a point, C, and a direction, $q_3 = (O - C)/||O - C||$, in k dimensions we need k–2 orthogonal directions, say $q_3,q_4, \ldots, q_k$. Two further directions, q_1 and q_2 define the viewing coordinates. Since the perpendicular to W in k dimensions is a (k–2)-space, O is any point of the form

$$O = \sum_{j=3}^{k} s_j q_j$$

for chosen scalars s_j. However, the equation of the plane remains (6), so that α still is given by (7), and x_s, y_s by (8), where the dot products are now of k-vectors.

There are many other ways to derive the perspective coordinates. The preceding has an appeal in data analysis, in that one frequently wants to choose W so as to display structure in a set of k-dimensional data. It then seems quite plausible that the orthogonal matrix Q may be derived from the analysis (for example as the matrix of principal directions in principal component analysis in Section **5.k**). Display of the data projected using chosen pairs of directions from Q will then be natural and, hopefully, informative.

Notice that as the distance, OC, increases, the line PP' becomes gradually parallel to OC, that is, perpendicular to W. This limiting case is important because it is simple and because the general case can be reduced to it. It is generally called *orthographic* or *parallel* projection. Each point, P, is projected perpendicularly onto W; in terms of our k-dimensional formulation, P is replaced by its components along q_1 and q_2.

We mentioned *homogeneous coordinates* in Section **c**. For a point, P, in k dimensions, we represent P by

$$x = (x_1, \cdots, x_{k+1})$$

with the understanding that $x_{k+1} \neq 0$ and that Cx is equivalent to x for any scalar $C \neq 0$. Several simplifications result from this form, one being the reduction of perspective projection to parallel projection. In Figure 3, with $k = 3$, suppose we take W to be the plane $x_3 = 0$, let C be the origin, (0,0,0,1) and O the point (0,0,-d,1). It can be shown that the perspective projection of P in Figure 3 is the same point as the parallel projection of the point $A \cdot x$, in homogeneous coordinates, where

$$A = \begin{bmatrix} 1 & 0 & 0 & 0 \\ 0 & 1 & 0 & 0 \\ 0 & 0 & 1 & d^{-1} \\ 0 & 0 & 0 & 1 \end{bmatrix} . \tag{9}$$

Thus we can operate on the data defining our display as above and then proceed via the simpler parallel projection. Note that A depends only on W and O, and that linearity is preserved by this transformation.

e. Plotting Curves

Intuitively, a curve in a space of two or more dimensions is a one-dimensional, generally nonlinear, subset of points. For most applications, the natural mathematical analog is a *parametric representation* of a curve by a vector-valued function, defined as follows:

> The function $c(t)$, where t is a real variable on an interval $[a,b]$ and $c(t)$ is a k-vector, is a parametric representation of a smooth curve in k dimensions if dc/dt is continuous and not equal to 0 on $[a,b]$.

The requirement that $dc/dt \neq 0$ ensures that the points generated by $c(t)$ do not pile up at any t. The definition extends to *piecewise* smooth curves (curves with corners or cusps) by letting dc/dt have a finite number of discontinuities. Parametric representations are not the only useful representation; occasionally one may leave the curve expressed in an equation form such as $x^2 + y^2 = r^2$, rather than $(r\cos(t), r\sin(t))$. In our discussion, however, we emphasize the parametric form as being the most general and useful.

The methods discussed in Section **d** for perspective projection allow us to translate points on curves in k-space into a projection on the two-dimensional viewing plane. Therefore, we specialize for the moment to the case of planar curves $(x(t), y(t))$. Computationally, such curves will be provided either by an algorithm or by a set of data points (x_i, y_i). Consider the latter situation.

In drawing the curve, the simplest approach is to join the points in order by straight-line segments. This method is recommended when the set of points is quite dense, when there is no supposition that the underlying curve is smooth, or when it is crucial to avoid any implications about the shape of the curve not directly provided by the data. One expects these considerations to be important in analytical study of observed data. In addition the simplest procedure is also the fastest and cheapest, at least in processing time, so that line-segment plotting will have advantages for interactive or high-volume displays.

If a smoother, more attractive curve is desired, one needs to derive a smoothing/interpolating curve based on the given points. For computational simplicity and control over graphical quality, the derived curve should be expressed in parametric form. A simple technique is to fit smoothing splines, $x(t)$ and $y(t)$, to the given data points, using techniques described in Section **4.e**.

A different algorithm, given by McConalogue (1970; 1971) also generates a smooth piecewise polynomial (here cubic) representation $(x(t), y(t))$, but making use of tangent information about the original curve. If the angle to the x-axis made by a tangent to the curve at (x_i,y_i) is known, say θ_i, the algorithm fits cubics $p_i(t)$, $q_i(t)$ on an interval $[0,t_i]$ from (x_i,y_i) to (x_{i+1},y_{i+1}) such that (p_i,q_i) agrees with (x_i,y_i) at $t = 0$ and (x_{i+1},y_{i+1}) at $t = t_i$, in value and in direction; specifically,

$$dp_i/dt = \begin{cases} \cos\theta_i & \text{at } t = 0 \\ \cos\theta_{i+1} & \text{at } t = t_i \end{cases} \qquad (10)$$

$$dq_i/dt = \begin{cases} \sin\theta_i & \text{at } t = 0 \\ \sin\theta_{i+1} & \text{at } t = t_i. \end{cases}$$

Once a parametric form is chosen there remains the question of drawing the curve. This will usually involve a sequence of line segments connecting points computed along the approximating curve. One criterion for choice of these points is that the *arc length* between them be constant. The arc length, $L(t_*)$, from $t = 0$ to $t = t_*$ of the approximating curve, $h(t) = (p(t), q(t))$, is

$$L(t_*) = \int_0^{t_*} ||dh/dt||\, dt.$$

This defines $L(t)$ as a function of t in each subinterval. The algorithm of McConalogue (1971) makes $L(t)$ approximately equal to t, so that equally spaced points in t (the easiest way to draw the curve) give good arc lengths. Note that $dL/dt = 1$ at $t = 0$ and $t = t_i$ from (10). Also, t_i is chosen to make $dL/dt = 1$ at $t_i/2$. See the references for details and a FORTRAN algorithm.

When no slopes are available, one may use techniques for approximating derivatives. For example, the spline interpolation procedures give good approximate derivatives. McConalogue (1971) uses instead the derivative of a three-point parabolic approximation. Of course, direct spline approximation of $x(t)$ and $y(t)$ avoids estimating slopes, although one then should consider the question of arc length in choosing the points to be plotted from $h(t)$.

f. Plotting Surfaces; Hidden Line Removal

A surface in k dimensions, in its most general form, is a two-dimensional nonlinear subset. By analogy with curves, the parametric representation of a smooth surface is a continuously differentiable vector-valued function of two variables, $s(t,u)$, such that neither $\partial s/\partial t$ nor $\partial s/\partial u$ vanishes. As with curves, the surface may be defined by an algorithm or a set of points. In general, the latter would consist of a set of parameter points, (t_i, u_i) and the corresponding k-dimensional points, s_i, representing the surface. A convenient specialization is to a *regular* grid of values (t_i, u_j), for $1 \leqslant i \leqslant m$ and $1 \leqslant j \leqslant n$, and the corresponding points, s_{ij}. Notice that, for the most general surface, t and u will not be ordinary co-ordinates, but special parameters. Consider, for example, parametrizing a sphere in this form: t and u might be angles in three-dimensional polar coordinates.

For purposes of plotting, several specializations occur. Most procedures start from a surface in three dimensions, so that the function s and the points s_{ij} are three dimensional. Often, the discussion is further specialized to take t and u to be the coordinate axes, say x and y. Provided that the surface is single-valued, it may then be reduced to a function, $z(x,y)$, which gives the z coordinate as a real-valued function of x and y. Similarly, the representation as a set of points becomes a matrix Z such that $z_{ij} = z(x_i, y_j)$. Whether the specialization to single-valued surfaces is taken has a major effect on the kind of algorithm used for plotting. Applications such as plotting of likelihoods or other functions of two variables naturally occur in this form. However, for other applications, such as plotting spheres, ellipsoids or other objects, one must, at best, break the surface into several single-valued functions. For complex objects, the representation in this form may be very inconvenient. Before using one of the procedures discussed in this and the next section, one should make sure that the representation used is reasonable for the application.

There are two fundamentally different approaches to plotting the surface, $s(t,u)$, according to whether curves or faces are displayed. The former represents the surface by the curves generated from S; i.e., by the functions, $s(t,u_j)$ and $s(t_i,u)$, of t and u, defined by fixing the other parameter. The latter plots the *face* defined by the region, $(t_i \leqslant t \leqslant t_{i+1}, u_j \leqslant u \leqslant u_{j+1})$, as a (usually) opaque region on the viewing window, which will be colored or shaded appropriately. While producing more realistic and attractive pictures, the face display is much more expensive and time-consuming. Most data-analytic applications do not require or justify this additional work and will rely on the curve-drawn procedures. Readers interested in the more elegant displays may consult Newman and Sproull (1973) or the survey paper by Sutherland et al. (1974).

Given a choice of grid and a perspective projection, surface plotting then reduces to two-dimensional curve plotting. After the points on the projected curves have been determined, the smoothing and curve-plotting procedures above may be applied. Alternatively, one could generalize the smoothing procedures to apply to surfaces, and then project the smoothed approximation. This involves more computation but might be preferred if several views of the surface were to be displayed.

One aspect of a more elegant display that may be of interest is the elimination of *hidden lines.* If a point on the surface lies directly behind another point on the surface, along the line to the observer, the farther point is *hidden.* For example, in Figure 3 of Section **d**, any point on the line OP will also project into P'. The elimination from a display of that part of a line or face hidden from the observer constitutes the hidden-line problem. A full treatment involves more sophisticated techniques than data analysis usually needs, but some moderately efficient simplified procedures are available.

Consider Figure 3 again, but now let us project perpendicularly down onto the plane containing OC and the x-axis in W, as in Figure 4. The viewing window projects into the line segment W_* and the points P and P' into P_* and P'_*, respectively. A simplification of hidden line removal results from noticing that all the points on the segment PP' (i.e., all the points hiding P) project down into $P_*P'_*$.

This observation can be combined with the special choice of axes and the transformation (10) of **d**, so that points along $P_*P'_*$ are ordered by decreasing z coordinate.

With or without these simplifications, the algorithm must now generate the set of curves defining the surface, examine them to determine the range of visibility, and plot the visible portions. To do this efficiently, the component parts of the surface should be ordered so that the current portion cannot obscure a previously considered portion. Most modern algorithms have this feature in some form. There is great variety in the assumptions about the form of

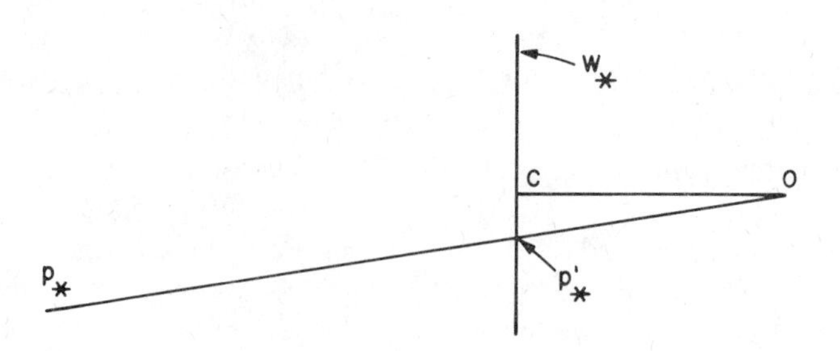

Figure 4: Hidden-Line Procedure

the surface and in the plotting technique. Many algorithms assume S to be a single-valued function of two coordinate axes.

Published algorithms in FORTRAN are given by Williamson (1972) and Wright (1974). The former plots the surface as a single-parameter family of curves; that is, plotting only one of the two sets of curves mentioned previously. This makes the sorting simpler and the algorithm faster, in the case of a pen plotter, but is not really necessary. Graham (1972) describes, without an algorithm, how to generate both sets of curves. Wright's algorithm draws essentially the successive curves formed by the intersection of the surface with each of a family of parallel planes. The result is nicely general in the type of object allowed, but specifying the surface in the required form may require considerable preprocessing for most data-analytic applications. (Also, the program is written in a particularly insidious nonstandard FORTRAN: users should note that all the entry points have implicitly the same arguments as the main entry.) The plots produced by this algorithm may also be startling to some users more accustomed to the curves or surface patches mentioned earlier. An unusual "wood-grain" effect is often produced.

A reasonably general algorithm producing an attractive plot for multiple-valued surfaces specified in a convenient form (say, as a set of single-valued surfaces) has not yet been published, to my knowledge. For further discussions of techniques in this field, see Sutherland et al. (1974).

g. Contour Plotting

Consider a single-valued function representing a surface in three dimensions, say $f(x,y)$. Typically, the surface will consist of either a computational algorithm able to generate the value of the surface given (x,y) or else a given set of points on the surface, for a specified set of (x_i,y_i). Strictly speaking, a *contour plot* attempts to draw the curves of constant value,

$$f(x,y) = v_k \tag{11}$$

for some chosen set of values, v_k, over a specified domain for (x,y). The term is also extended to plots which instead display intervals, $[v_k,v_{k+1}]$, of values for $f(x,y)$ by some type of coding. The information provided is nearly equivalent, and the choice of form derives largely from the type of display device. We will mention some slight differences in the visual effect of the two forms, however.

The simplest method of contour plotting assigns a character code to each interval and displays the appropriate character at each point at which the surface is given. The character codes should be chosen to give information about height; for example if there are ten or fewer intervals, the digits, 0,1,...,9, can be used. Another approach is to choose characters with increasing darkness or *grey level*. For a small number of levels, one can use ordinary characters; for example, dot, asterisk and overstruck 0 and X usually produce a legible three-level scale.

The technique so far described has three advantages: it is applicable to nongraphic devices such as printers; it accomodates itself to surfaces defined by an irregular set of points, (x_i,y_i), provided we remember not to use a blank as a level code; and it can produce fairly informative plots with only a few points. On the debit side, it produces rather coarse pictures and is poorly accommodated to representing a large amount of information, at least when the character size is fairly large. It is mostly useful for printer plots. As with other printer plots, one can write a reasonably portable program; surprisingly, there seems to be no general-purpose published algorithm for this simple procedure.

On *high-resolution* display devices, the grey-level process is capable of great refinement. The technique has mostly been used to reproduce digitized pictures, but it applies similarly to contour plotting. When the size of the individual plotting position is small, we are effectively coding *regions* rather than displaying single characters. We discuss this approach later in the section, since it is computationally more like line-drawn contours. Note that for very high resolution, one usually wishes to use basic plotting dots, rather than characters. It is interesting that the techniques involved date back again to the automatic loom of Jacquard in the early 19th century, the source of several essential concepts in computing. Knowlton and Harmon (1972) give a fascinating discussion of historical and modern grey-scale generation.

A third method of coding intervals is to use *colors* to represent the various levels. This technique is well known to any reader of atlases and maps in which height of land or depth of water is coded by the use of several colors. Precisely the same technique can be used to display contours of arbitrary functions. A suitable device is a fixed scan, color monitor of type 6 in Figure 1; in other words, a color television monitor. The color has the advantage that nonadjacent regions can be compared for height by noting the color. Also single intervals can be made to stand out throughout the display, again by a suitable choice of colors. By comparison with line-drawn contour plots, the up-or-down direction in small details can be easily determined from the colors. An interactive color contour-plotting system has been written by the author and used extensively at Bell Laboratories for some time (Chambers, 1973).

If we wish to draw the contour *lines* (11), as we would if working with one of the line-drawing displays of types 2, 4 or 5, we will need some numerical calculations to approximate points on these curves. We give first the simplest procedure and then discuss alternatives. Suppose we have a regular grid of points, (x_i,y_j), for $i = 1,...,m$, $j = 1,...,n$, and the array Z of corresponding values,

$$Z[i,j] = f(x_i,y_j) \; . \tag{12}$$

We now have a set of $(m-1)(n-1)$ adjacent rectangles, or *boxes*. Suppose we assume that $f(x,y)$ is *continuous*. Then we can examine each edge of each box to determine whether the contour line crosses that edge. Suppose (x,y) and (x',y') define an edge; that is, we have $(x,y) = (x_i,y_j)$ and either $(x',y') = (x_{i+1},y_j)$ or $(x',y') = (x_i,y_{j+1})$. Then the contour line (11) crosses the edge if

$$\min(z,z') \leqslant v_k \leqslant \max(z,z')$$

where $z = f(x,y)$ and $z' = f(x',y')$. If we examine each edge of a box, the contour line can cross zero, two or four edges by this criterion. The types of intersection are shown in Figure 5, with

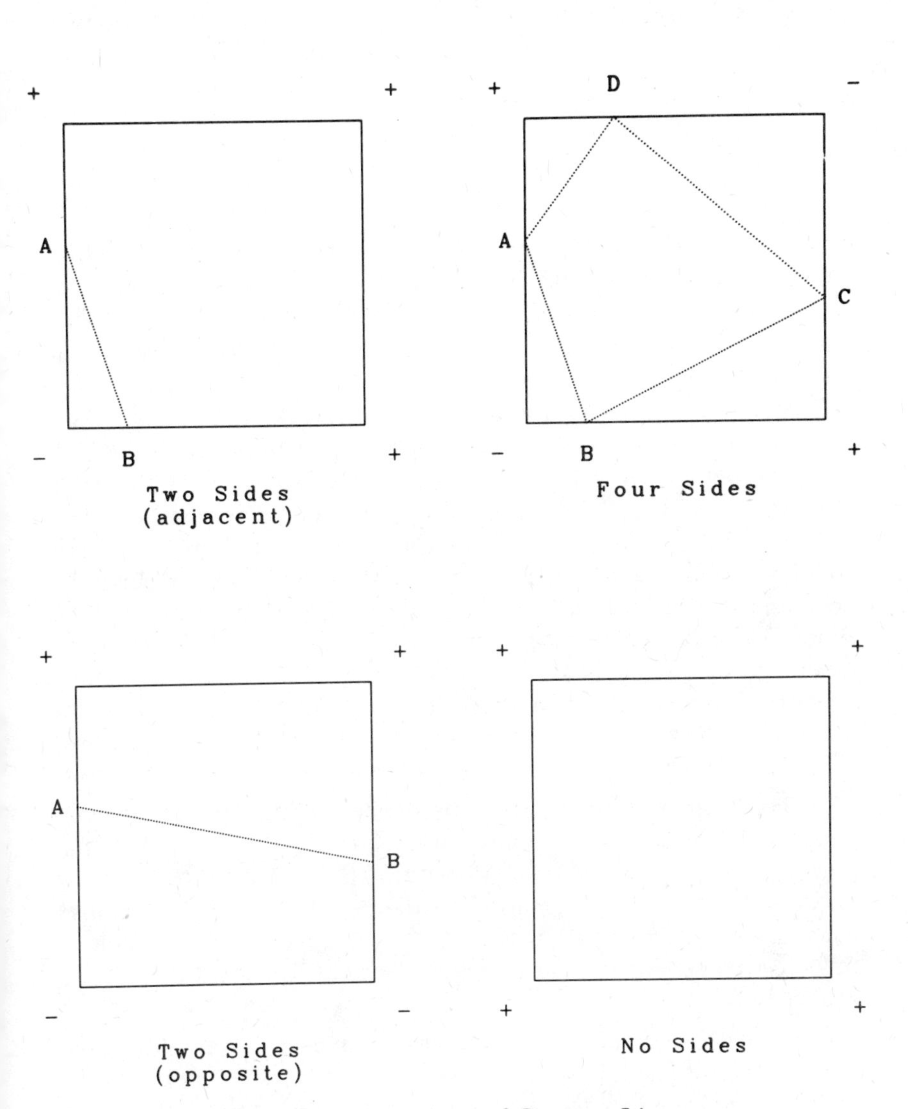

Figure 5: Intersections of Contour Lines

vertices marked + or − as $f(x_i,y_i)$ is larger or smaller than v_k. All other possibilities reduce to these by rotation, reflection and exchange of + and −. One now must estimate numerically the intersection points A, B, C and D, and draw the curve they define. The simplest method is to interpolate linearly and join the intersections by a straight line. In this form, the intersection point is

$$A = \alpha(x,y) + (1-\alpha)(x',y') \tag{13}$$

where

$$\alpha = (v_k - z')/(z - z') . \tag{14}$$

Notice that the 4-sided case is ambiguous, in that the lines could be AB and CD or AD and BC for the nonintersecting case, AC and BD for a saddle point or ABCDA for a closed local contour. Considering the function values of adjacent boxes will usually distinguish these cases.

In many situations, the speed of plotting contours is improved if the individual line segments are joined into a connected chain. For example, this reduces the amount of data which must be sent to the terminal and also the pen motion required on a pen plotter. The end points of all the line segments for a particular contour level are computed and saved. A postproccess then builds up a chain by matching the current ends of the chain to the remaining segments. When a match occurs the segment is added to the chain and the other end of the segment becomes the end of the chain. A simple version of this procedure is in the contour algorithm in Bell Laboratories (1977). A general and efficient procedure would make a useful published algorithm.

To obtain a smoother curve, one may use higher-order approximation and/or take the resulting curve and smooth it by one of the procedures in Section **e**. Note that the latter requires the chaining of the line segments. Note also that the points x_i and y_j need *not* be equally spaced. Generally a narrower spacing should be taken where the surface has sharp peaks or valleys, so that detail in the contours is needed.

Higher-order approximations have not been developed extensively. Marlow and Powell (1976) give a partial solution, with a FORTRAN algorithm, for quadratic approximation to the surface.

The reprecentation of contours by linear approximations on boxes extends to non-Euclidean coordinates; for example, we can express the surface in polar coordinates as $f(r,\theta)$. Boxes are now wedge-shaped, but the argument leading to the curve continues the same, computing the eventual plotting points in polar coordinates. Figure 6 shows a polar-coordinate plot of cylindrical shielding effects

Figure 6: Polar Co-ordinate Contour Plot

in cross-section (Wilson, 1974). The coordinates can be any system, provided that the box generated is a closed, convex curve. See Crane (1972) for a discussion of this process and an ALGOL 60 algorithm of the basic plot. The technique extends also to nonrectangular grids provided that the grid divides into contiguous, convex boxes.

The linearizing method is simple and general and is probably the best standard procedure. For some special functions the contours may be tracked more efficiently by an *incremental technique.* Given that f(x, y) is convex about a known center, C, we can track a contour using polar coordinates about C, solving for a grid of angles, θ_j, the equations

$$f(r_j,\theta_j) = v_k .$$

For general functions, however, an incremental solution for successive θ_j may not completely capture the curve. The previous algorithm is usually more successful, although it can never detect contours entirely inside one box and will also miss contour lines crossing one edge twice.

As noted, the computations for color, grey-level, or other coding of contour regions become computationally very similar to contour line drawing when the number of display points is large. The basic algorithm for the latter problem extends easily to the former. The display surface is divided into rectangular areas corresponding to the boxes in Figure 5. We assume $f(x,y)$ given at the vertices of the boxes, as before. Then the numerical approximation technique used to find the intersection points will also give an approximation to the function value at each display point inside the box. The color or other code appropriate for the approximate value is then displayed. For a point at $(x_i + \delta x,\ y_j + \delta y)$, linear interpolation approximates $f(x,y)$ by

$$z = \alpha\beta\, Z[i+1,j+1] + (1-\alpha)\beta Z[i,j+1] + \alpha(1-\beta)Z[i+1,j] + (1-\alpha)(1-\beta)Z[i,j]$$

where

$$\alpha = \delta x/(x_{i+1} - x_i)$$

$$\beta = \delta y/(y_{j+1} - y_j) .$$

More sophisticated approximations are possible, of course. For the sake of efficiency, one will avoid the arithmetic when possible: specifically, when the entire box is coded at one or possibly two levels (Chambers, 1973).

h. Scaling

The techniques described so far have concerned the plotting region, in the terminology of Section **a**. Now we consider some techniques for embedding the plot in an attractive and informative figure.

Given a set of data, one must choose lower and upper limits for user coordinates, and usually a set of axis labels to display on the axes. The appropriate algorithm depends somewhat on the type of plot, but generally is somewhat as follows. One set of coordinates for a given set of data determine minimum and maximum values, say $[u_a, u_b]$. One must then choose a user coordinate range, $[u_A, u_B]$, with

$$u_A \leqslant u_a \quad \text{and} \quad u_b \leqslant u_B \tag{15}$$

and a set of labeling points, $u_1, \ldots, u_k$ for numerical labels. The labeling points should be *simple,* in the sense that u_j can be represented as a decimal number with as few digits as possible, but the user coordinate range should cover the data range with as little wasted space as possible.

The following procedure usually produces an adequate range and labels, given u_a, u_b and a suggested value k_0, for k.

I. Choose a simple value, δ, such that $(k_0 - 1)\delta$ covers the width, $|u_b - u_a|$.

II. Find the largest $u_A \leqslant u_a$ and smallest $u_B \geqslant u_b$, among integer multiples of δ.

III. The labeling points are the sequence $u_A + (j-1)\delta$, for $j = 1,\ldots,k$ and $u_k = u_B$.

Notice that k may be larger than k_0. Usually, δ is taken to be one, two or five times a power of ten. Lewart (1973) gives a FORTRAN algorithm.

Some simple modifications may be useful. If one does not wish to waste the space resulting from expanding the data interval, one can take the user coordinate range to be the data range and choose u_j as *internal* axis labels. The algorithm above can be adapted to this case, by choosing u_a as the maximum and u_b as the minimum. Steps I and II are as before. The labeling points are now $u_B + (j-1)\delta$ for $j = 1,\ldots,k$, $u_A = u_k$ and $k \leqslant k_0$.

This approach will result in some of the data points lying exactly on the limits, u_a and u_b. The first algorithm may also give $u_A = u_a$ or $u_B = u_b$. If this is undesirable (because the plotted point would be obscured by the axis), the corresponding user coordinate limit should be extended, say by a character width, to clear the point. The labeling points remain as computed.

Notice that the minimum and maximum of the data are not *robust* estimates of the spread of the data, in the statistical sense. If one outlying value becomes very large, in either direction, all the remaining data are squeezed into a small subinterval and the plot becomes unreadable. Two steps should be taken to avert this difficulty. A robust estimate of the range should be made, and some form of rescaling applied to any points outside this range.

The following simple range estimate is adequate for most cases. We compute the order statistics $X_*[N\alpha]$ and $X_*[N(1-\alpha)]$ (see Section **3.d**), where we expect fewer than α fraction of outliers at either end. Typically, one takes $\alpha = .25$ and computes the quartiles. Then a range estimate is

$$u'_a = (1+c)X_*[N\alpha] - cX_*[N(1-\alpha)]$$

$$u'_b = -cX_*[N\alpha] + (1+c)X_*[N(1-\alpha)]$$

where c is some chosen constant. For data from a given distribution, the two limits above approach, as $N \rightarrow \infty$, q_a and q_b, the quantiles of the distribution obtained by setting $X_*[N\alpha]$ and $X_*[N(1-\alpha]$ to the αth and $(1-\alpha)^{th}$ quartile of the distribution. By computing the tail probabilities for q_a and q_b, the behavior of a particular rule can be determined. For example, $c = 1.5$ would exclude about about 1% of a normal sample. Given u'_a and u'_b, the robust range is then taken to be

$$[\max(u_a, u'_a), \min(u_b, u'_b)]$$

with $[u_a, u_b]$ the raw data range. The following section includes an example of rescaling.

Special applications may suggest alternative scaling algorithms. When the axis represents *time* in plots of economic or similar data, the labeling points will normally be chosen in terms of the basic period of the data (for example, months or quarters), rather than on a decimal scale. An algorithm is included in the Bell Laboratories (1977) package.

For histograms, the number of bars is chosen by an argument related to scaling. See Section **j**.

i. Scatter Plots

Given n observations on two variables, x and y, we plot a character or symbol at the n points representing (x_i, y_i). The straightforward implementation chooses a plotting interval for the x axis from $[\min_i x_i, \max_i x_i]$, as discussed in Section **h**, and similarly for y. The basic geometry discussed in Section **c** then defines the plot.

The scatter plot is the workhorse of statistical graphics. Even in its simplest form, it provides the essential tool for displaying results. A few additional facilities and some care in designing the algorithm, as discussed in Section **a**, make the plot more useful. We consider next robust intervals for scatter plots and the treatment of coincident points.

As noted in **h**, the intervals based on the full range of x and y are not robust to outlying values. A robust interval may be chosen as described there, which may not contain all the observed values. A scatter plot using robust intervals may simply omit such points (printing a warning message and the omitted observations). A preferable strategy is to plot rescaled versions of the outlying points as follows. Let $[u'_A, u'_B]$ be the robust interval and $[u_{min}, u_{max}]$ the full range. If outliers exist on either end, we choose to allow at most a fraction α of the inner range to plot outliers on that end. This results in a full range $[u''_A, u''_B]$ where

$$u''_A = \max \begin{cases} u_{min} \\ (1+\alpha)u'_A - \alpha u'_B \end{cases}$$

and

$$u''_B = \min \begin{cases} u_{max} \\ (1+\alpha)u'_B - \alpha u'_A \end{cases}$$

Then if, for example, $u < u'_A$, one plots instead $u' = (1-\beta)u'_A + \beta u''_A$, with $\beta = (u-u_{min})/(u'_A - u_{min})$. A similar rescaling on the other end occurs if $u > u'_B$. The robust scatter plot rescales both x and y, when necessary, and plots the rescaled points. For further analysis, one should also label these points or print them out, or both. The outlying regions on the plot, if any, should be marked off by grid lines drawn at x'_A, x'_B, y'_A or y'_B. Scatter plots based on similar ideas were developed by H. J. Chen at Bell Laboratories. Another version is included in the Bell Laboratories (1977) graphics package.

If two or more observations in a scatter plot are identical or nearly so the plotted symbols will coincide, masking possibly important information. The natural solution is to plot at the point a count of the number of observations, along with or in place of the usual symbol. A reasonably efficient procedure to compute the counts is as follows.

The complete plotting region is imagined as a grid of squares, such that all points in a given square are treated as coincident. (The size of the square should presumably be a function of the size of the displayed symbol.) Each observation corresponds to a cell, say (j, k) from the J by K grid, which can be coded as a single value, say $c = j + (k-1)J$. Then one sorts the values, c_i, corresponding to the observations, carrying the labels, i, along as a passive array. Coincident points in the sorted data are then counted and the count plotted, say at (x_i, y_i) for one of the observations corresponding.

j. Histograms and Probability Plots.

The simplest *histogram* consists of a set of M bars, the j-th bar being defined as the rectangle $(x_j \leqslant x \leqslant x_{j+1}, 0 \leqslant y \leqslant h_j)$. On a line-drawing device, there are several algorithms for plotting these bars compactly. One is to plot the horizontal line segments

$$(x_j,h_j) \text{ to } (x_{j+1},h_j)$$

and the vertical line segments

$$(x_j,\max(h_{j-1},h_j)) \text{ to } (x_j,0).$$

Here, we formally set $h_0 = 0$. On variable scan displays or pen-type plotters, one will want to reduce the excess movement of the plotting point. One such procedure is to plot first the connected line segments which define the top of the histogram; namely,

$$(x_1, h_0),\ (x_1, h_1),\ (x_2, h_1),\ (x_2, h_2),\ \cdots$$

and then, optionally, drop verticals to the axis.

For a general-purpose histogram plotting routine, the user may prefer to provide directly a set of data points which will be grouped automatically by the histogram routine rather than specify the precise bar positions and heights. Both options should be available. A common choice for the number of bars corresponding to N points is

$$M = 1 + \log_2(n)$$

Doane (1976) proposes adding to the above when the distribution is not symmetric; specifically,

$$M = 1 + \log_2(n) + \log_2(1+\beta)$$

where β is a standardized measure of skewness,

$$\beta = \frac{m_3}{m_2^{3/2}} \left(\frac{(n+1)(n+3)}{6(n-2)} \right)^{1/2}$$

with $m_j = n^{-1} \sum_i (x_i - \bar{x})^j$. Given the number of bars, neat values for the break points may be chosen, by a method such as described in Section **h**. Doane recommends trying several values of M to get the best result.

There are several special possibilities and considerations in plotting histograms. Changes of scale, of orientation and of cell shape may be made to emphasize statistical or other properties of the data (Tukey, 1977). When the data contain outliers, the method of computing a robust interval and rescaling, discussed in **h**, may be used.

Probability Plots. The *probability* or more specifically *quantile-quantile* plot may be defined as the plot of two inverse distribution or quantile functions, $Q_1(p)$ and $Q_2(p)$, for $0 < p < 1$. One plots the points (Q_1, Q_2) corresponding to chosen values of p.

The most common cases are that one or both of Q_1 and Q_2 are defined by empirical distributions; i.e., by a set of observations. The empirical-theoretical plot takes, say, observations $y_1, \ldots, y_n$, orders them and plots the ordered y_i against the corresponding quantile, $x_i = Q(p_i)$, of some chosen reference distribution (such as the standard normal). The fraction p_i may be considered as the approximate fraction of the reference distribution which should be less than y_i. A value of $p_i = (i+.5)/(n+1)$ is a good rule of thumb. A possible optimal choice of p_i has generated some papers in the literature, but will make no practical difference for sensible sample sizes.

In an empirical-empirical plot, both axes represent empirical distributions. If the two samples are of equal size, this is merely a scatter plot of one set of sorted values against the other. If the samples are of different sizes, one may interpolate in the larger sample to estimate a value of $Q(p_i)$ corresponding to each p_i in the smaller sample.

Statistical applications of probability plots are described in Wilk and Gnanadesikan (1968). The essential property of any form of probability plot is that, if the two distribution functions are identical the plot will approximate a line with unit slope and zero intercept; i.e., $y = x$. Some applications of the plots are:

1. *Comparison of distributions.* If the data were generated from a random sample of the reference distribution, the plot should look roughly linear. This remains true if the data come from a linear transformation of the distribution (see 3.).
2. *Outliers.* If one or a few of the data values are contaminated by gross error or are for any reason markedly different in value from the remaining values, and the remaining values are more or less distributed like the reference distribution, the outlying points may be easily identified on the plot.
3. *Location and Scale.* Because a change of one of the distributions by a linear transformation simply transforms the plot by the same transformation, one may estimate graphically location and scale parameters for a sample of data, on the assumption that the data comes from the reference distribution. It is possible in such graphical estimation to take into account outliers, etc., in an informal way.
4. *Shape.* Some differences in distributional shape may be deduced from the plot. For example, if the y-axis distribution has larger tails (tends to have more large values) the plot will curve down at the left and/or up at the right.

Probability plots are an important data-analytic tool, providing many insights but also presenting some interpretive difficulties. It is important to remember that the variance of the order statistics generally increases as one moves to the ends of the sample. A confidence interval may be computed for an individual point and shown, for example, by a vertical bar. Definition of useful joint confidence regions remains an unsolved problem.

There are two aspects to probability plotting from the viewpoint of computations: the numerical computation of quantiles, and the graphical display of the results. Numerical approximation and applications to probability distributions were discussed in Chapters 4 and 7 above. The present case has one notable special property; namely, one wishes to approximate a *set* of values for the given distribution. There is a possibility of using this fact (for example in an iterative computation of Q(p)) to reduce the time taken per value over a standard algorithm. In so far as the actual plotting is concerned, the primary need is for a flexible scatter plotting procedure of the form described above. In fact, all the requirements mentioned might prove useful in some circumstance for probability plots.

It will generally be desirable to create a specialized set of probability-plotting algorithms. Common distributions (normal, Cauchy, gamma, etc.) should ideally be offered as separate algorithms, along with facilities for general insertion of an arbitrary empirical or theoretical (numerical) distribution.

Because the distributions in the probability plot must be completely specified, up to location and scale, empirical probability plots against general families of distributions require the estimation of unknown parameter values. For example, to plot against the gamma (chi-square) distribution, the shape parameter (degrees of freedom) must either be known or estimated from the sample by, say, maximum likelihood. General techniques for estimation are discussed in Chapter 6. A number of ad-hoc techniques are also used for particular distributions (Wilk and Gnanadesikan, 1968). Further research is needed, particularly on estimation procedures more closely related to the plotting. Robust estimates are desirable (compare Section **5.j**). In an interactive graphical facility, one may want to adjust the parameters and replot, based on looking at the current plot.

k. Summary and Recommendations

In contrast to some numerical applications, algorithms for graphics gain a great deal from being developed into a coordinated general system rather than operating separately for each problem. In Section **a** some general concepts in the design of such a system are presented, related to the Bell Laboratories (1977) system for graphics in data analysis. In evaluating systems of graphical operations, the special needs of data analysis should be considered. Some specific points include:

(i) *Operations provided.* Does the system provide suitable operations for common data-analytic needs, such as scatter plots and histograms?

(ii) *Flexibility of control.* Systems should provide both high-level operations and lower-level control to construct or modify these operations. This is particularly important since data analysts may need new graphical operations on a system not designed primarily for them.

(iii) *Devices.* Does effective use of the system assume expensive equipment, either in the terminal or the supporting computer? Is it possible and easy to generate output on different devices? Does the system adjust its internal strategy to make good use of both crude and high-quality devices?

(iv) *Complexity.* How much of the system's facilities are actually relevant to data analysis? Do elaborate data structures or program facilities impose a serious penalty on simple applications in cost, response time, or program size?

Separate algorithms for some specific graphical operations are available. Crane (1972) gives a contour plotting algorithm in ALGOL60; Williamson (1972) and Wright (1974) give algorithms in FORTRAN for plotting surfaces with hidden lines removed; Lewart (1973) gives an algorithm for the choice of axis limits and Doane (1976) a prescription for limits on histograms.

Problems.

1. Describe how a program for drawing a contour plot, as in Section **g**, could be adapted to start from an arbitrary set of points, (x_i, y_i, z_i), rather than the values of z at a *regular* grid of x and y values. Include the possibility that the values, z_i, include random error.

2. Given a method for choosing an interval and labeling points as in Section **h**, design a program to solve the following problem. Given m input intervals, $[u_a^{(j)}, u_b^{(j)}]$, $j=1, \cdots, m$, construct m output intervals, $[u_A^{(j)}, u_B^{(j)}]$, with corresponding labeling positions, such that all the intervals are of equal width. (This is useful, for example, in plotting several sets of data on comparable scales.)

3. Write an algorithm to shade in a bar with parallel lines. The arguments should include the bar (say the lower-left corner, width and height), the slope of the lines, the perpendicular distances between them and the coordinates of a fixed, but arbitrary, point on one of the lines. To see the value of the last argument, consider shading in all the bars in a histogram, with the shading lines passing continuously from one bar to the next. (Based on unpublished work of R. A. Becker and P. Guarino.)

4. Discuss the advantages and disadvantages of various methods of enhancing a scatter plot to display one, two, or more variables in addition to x and y. Some possibilities include: characters, drawn objects of varying size or shape, and shading.

REFERENCES

Note: The pages in the text on which the reference is cited are listed in italics after the reference.

Chapter 2

Aho, A. V. and Johnson, S. C. (1974). LR parsing. *Computing Surveys,* **6**, 99-124. *22*

Brown, W. S. (1977). *ALTRAN User's Manual.* Bell Laboratories, Murray Hill, NJ. *20*

Chambers, J. M. (1971). Algorithm 410: Partial sorting. *Comm. A. C. M.* **14**, 357-358. *12*

Dahl, O. J., Dijkstra, E. W., and Hoare, C. A. R. (1972). *Structured Programming.* Academic, New York. *15*

Fox, P., Hall, A. D., and Schryer, N. L. (1976). *The PORT Mathematical Subroutine Library.* Bell Laboratories, Murray Hill, NJ. *20*

Hall, A. D. (1972). *ALTRAN Installation and Maintenance.* Bell Laboratories, Murray Hill, NJ. *20*

Hoare, C. A. R. (1971). Proof of a program: FIND. *Comm. A. C. M.* **14**, 39-45. *17*

Johnson, S. C. (1976). Compiler-compilers: Where have we been and where are we doing? *Proc. Ninth Interface Symposium on Statistics and Computing* 56-58, Prindle, Weber and Schmidt, Boston. *.22*

Kernighan, B. W. and Plauger, P. J. (1974). *Elements of Programming Style.* McGraw-Hill, New York. *14*

Kernighan, B. W. and Plauger, P. J. (1976). *Software Tools.* Addison-Wesley, Reading MA. *14 26 27*

Knuth, D. E. (1974). Structured programming with **goto** statements. *Computing Surveys,* **6**, 261-302. *19*

Knuth, D. E. and Pardo, L. T. (1976). The early development of programming languages. *Tech. Rept. Stanford Univ. Dept of Computer Science.* (to appear in *Encylopedia of Computer Science and Technology). 21*

Knuth, D. E. and Stevenson, F. R. (1973). Optimal measurement points for program frequency counts. *BIT* **13**, 313-322. *13*

Marcotty, M., Ledgard, H. F., and Bochmann, G. V. (1976) A sampler of formal definitions. *Computing Surveys* **8**, 191-276. *22*

Mills, H. D.(1973). How to Write Correct Programs and Know it. IBM Corporation, Gaitesburg, MD. *15*

Ryder, B. G. (1974). The PFORT Verifier. *Software — Practice & Experience.* **4**, 359-377. *20*

Sammet, J. E. (1969). *Programming Languages: History and Fundamentals.* Prentice-Hall, Englewood Cliffs, NJ. *23*

Sande, G. (1975). Program execution profiles. *Proc. Eighth Interface Symposium on Statistics and Computing.* Health Sciences Computing Facility, UCLA, Los Angeles, CA.

Tannenbaum, A. S. (1976). A tutorial on ALGOL68. *Computing Surveys,* **8**, 155-190. *22*

von Neumann, J. and Goldstine, H. H. (1947). Numerical inverting of matrices of high order. *Bull. Amer. Math. Soc.,* **53**, 1021-1099. *21*

Chapter 3

Baecker, H. D. (1970). Implementing the ALGOL68 heap. *B.I.T.* **10**, 405-414. *32*

Bray, D. W. (1974). Dynamic storage routines for FORTRAN programs. *SIGPLAN Notices* September, 1974. *32 55*

Brent, R. P. (1973). Reducing the retrieval time of scatter storage techniques. *Comm. A. C. M.* **16**, 105-109. *49 51 56*

Chambers, J. M. (1971). Algorithm 410: Partial sorting. *Comm. A. C. M.* **14**, 357-358. *45 55*

CODASYL (1971). *Data Base Task Group Report.* Assoc. Comp. Mach. *38*

Date, C. J. (1975). *An Introduction to Data Base Systems.* Addison-Wesley, Reading MA. *38 40*

Fox, P. A., Hall, A. D. and Schryer, N. L. (1975). Basic utilities for portable FORTRAN libraries. *Comp. Sci. Tech. Report No. 37* Bell Telephone Laboratories, Murray Hill. *30 55*

Hoare, C. A. R. (1961). Algorithm 63: Partition; Algorithm 64: Quicksort. *Comm. A. C. M.* **4**, 321. *42*

Knuth, Donald E. (1968). *The Art of Computer Programming* **1**, (Fundamental Algorithms) Addison Wesley, Reading MA. *32 35*

Knuth, Donald E. (1973). *The Art of Computer Programming* **3**, (Sorting and Searching) Addison Wesley, Reading MA. *44 45 47 49 50 51 52 55 56*

Loesser, R. (1974). Some performance tests of "quicksort" and descendants. *Comm. A. C. M.* **17**, 143-152. *44*

Loesser, R. (1976). Survey on Algorithms 347, 426, and Quicksort. Trans. on Math. Software, **2**, 290-299. *44*

Lorin, H. (1975). *Sorting and Sort Systems.* Addison-Wesley. *44 45 55*

M.I.T. Computation Center (1962). *LISP 1.5 User's Manual. 36*

Millstein, R. E. (1973). Control structures in Illiac IV Fortran. **16**, 621-627. *34*

Payne, W. H. (1973). Partial sorting: a large vector technique and its applications. *Int. J. Comp. and Inf. Sci.* **2**, 141-156. *45*

Singleton, R. C. (1969). Algorithm 347: Sort. *Comm. A. C. M.* **12**, 185-186. *43 44 55*

Chapter 4

Ahlberg, J. H., Nilson, E. N. and Walsh, J. L. (1967). *The Theory of Splines and Their Applications* Academic, New York. *77*

Barrodale, I., Powell, M. J. D. and Roberts, D. K. (1972). The differential correction algorithm for rational l-infinity approximation. *SIAM J. Numer. Anal.* **9**, 493-504. *76 77*

Bergland, G. D. (1969). A guided tour of the fast Fourier transform. *IEEE Spectrum* **6**, 41-52. *96*

Bloomfield, P. (1976). Fourier Analysis of Time Series: An Introduction. John Wiley and Sons, New York. *91 96*

Blue, J. L. (1975). Automatic numerical quadrature: DQUAD *Comp. Sci. Tech. Report No. 25,* Bell Laboratories, Murray Hill, NJ.

Blue, J. L. (1977). A portable FORTRAN program to find the euclidean norm of a vector. *Trans. on Math. Software,* (to appear). *99*

Braess, D. (1971). Chebyshev approximation by spline functions with free knots. *Numer. Math.* **17**, 357-366. *81*

Brigham, E. O. (1974). *The Fast Fourier Transform.* Prentice-Hall, Englewood Cliffs. *91*

Brown, W. S. and Richman, P. (1969). The choice of base. *Comm. A. C. M.* **12**, 560-566. *61*

Clenshaw, C. W. (1962). *Chebyshev Series for Mathematical Functions.* National Physical Laboratory Mathematical Tables **5**, London. *83*

Cody, W. J., Fraser, W. and Hart, J. F. (1968). Rational Chebyshev approximation using linear equations. *Numer. Math.* **12**, 242-251. *75*

Cooley, J. W. and Tukey, J. W. (1965). An algorithm for machine calculation of complex Fourier series. *Math. Comp.* **19**, 297-301. *92*

Cox, M. G. (1971). An algorithm for approximating convex functions by means of first degree splines. *Computer J.* **14**, 272-273. *81*

de Boor, C. (1972). On calculating with B-splines. *J. Approx. Theory,* **6**, 50-62. *78 86*

Feller, W. (1971). *An Introduction to Probability Theory and its Applications.* John Wiley and Sons, New York. *91*

Gentleman, W. M. (1972). Implementing Clenshaw-Curtis Quadrature, I Methodology and experience; II Computing the cosine transformation. *Comm. A. C. M.* **15**, 3437-346. *94*

Gentleman, W. M. (1972a) Algorithm 424. Clenshaw-Curtis quadrature. *Comm. A. C. M.* **15**, 353-355.

Gentleman, W. M. and Sande, G. (1966). Fast Fourier transforms: for fun and profit. *Proc. Fall Joint Comp. Conf.* **29**, 563-578. *93*

Golub, G. H. and Smith, L. B. (1971). Chebyshev approximation of continuous functions by a Chebyshev system of functions. *Comm. A. C. M.* **14**, 737-746. *76*

Hart, J. F., Cheney, E. W., Lawson, C. L., Maehly, H. J., Mesztenyi, C. K., Rice, J. R., Thacher, H. C., and Witzgall, C. (1968). *Computer Approximations.* John Wiley and Sons, New York. *76 84 85*

Heller, D. (1976). A survey of parallel algorithms in numerical linear algebra. *Tech. Report AD-A024 792.* Carnegie-Mellon University, Department of Computer Science, Pittsburgh, PA. *99*

Herriot, J. G. and Rheinsch, H. (1973). Algorithm 472: Procedures for natural spline interpolation. *Comm. A. C. M.,* **16**, 763-768. *79*

Hildebrand, F. B. (1956). *Introduction to Numerical Analysis.* McGraw-Hill, New York. *88*

Jenkins, G. M. and Watts, D. G. (1968). *Spectral Analysis and its Applications.* Holden-Day, San Francisco. *91 96*

Kaneko, T. and Liu, B. (1970). Accumulation of round-off error in fast Fourier transforms. *J. A. C. M.* **17**, 637-654. *93*

Knuth, Donald E. (1969). *The Art of Computer Programming* **2**, (Semi-numerical Algorithms) Addison Wesley, Reading MA. *60*

Powell, M. J. D. (1967). On the maximum errors of polynomial approximations defined by interpolation and by least squares criteria. *Computer J.* **9**, 404-407. *76 89*

Rabiner, L. R., Cooley, J. W., Helms, H. D. Jackson, L. B., Kaiser, J. F., Rader, C. M., Schafer, R. W., Steiglitz, K. and Weinstein, C. J. (1972). Terminology in digital signal processing. *IEEE Trans. Audio and Electroacoustics,* **20**, 322-337. *92*

Ralston, A. (1967). *A First Course in Numerical Analysis.* McGraw Hill, New York. *74*

Rice, J. R. (1964). *The Approximation of Functions.* Volume 1: Linear Theory. Addison-Wesley, Reading Mass. *74*

Rice, J. R. (1969). *The Approximation of Functions.* Volume 2: Non-linear and Multivariate Theory. Addison-Wesley, Reading, Mass. *74 76 77 78 81*

Schumaker, L. (1968). Uniform approximation by Chebyshev spline functions, II: free knots. *SIAM J. Numer. Anal.* **5**, 647-656. *81*

Singleton, R. C. (1968). Algorithm 338: ALGOL procedures for the fast fourier transform; Algorithm 339: An ALGOL procedure for the fast fourier transform with arbitrary factors. *Comm. A. C. M.* **11**, 773-779. *93 94*

Singleton, R. C. (1969). An algorithm for computing the mixed radix fast Fourier transform. *IEEE Trans. Audio and Electroacoustics.* **17**, 93-103. *93 94 95*

Sterbenz, P. H. (1974). *Floating-Point Computation.* Prentice-Hall, Englewood Cliffs. *66*

Stewart, G. W. (1973). *Introduction to Matrix Computation.* Academic, New York. *71*

Werner, H., Stoer, J. and Bommas, W. (1967). Rational Chebyshev approximation. *Numer. Math.* **10**, 89-306. *75*

Wilkinson, J. H. (1963). *Rounding Error in Algebraic Processes.* Prentice-Hall, Englewood Cliffs. *64 66*

Woodford, C. H. (1970). An algorithm for data smoothing using spline functions. *B.I.T.* **10**, 501-510. *80*

Chapter 5

Note: A number of the ALGOL procedures are reprinted in Wilkinson and Reinsch (1971). These are noted in the references with *HAC(II)* and page numbers.

Andrews, D. F.(1974). A robust method for linear regression. *Technometrics* **16**, 523-531. *124*

Barrodale, I. and Roberts, F. D. K. (1974). Algorithm 478: Solution of an overdetermined system of equations in the L1 norm. *Comm. A. C. M.,* **17**, 319-320. *125*

Bartels, R. and Conn, A. (1978). Linearly constrained discrete L1 problems. *Trans. on Math. Software,* (to appear). *125*

Björck, A. (1967). Solving linear least squares problems by Gram-Schmidt orthonormalization, *B.I.T.* **7**, 1-21. *105 119*

Björck, A. (1967a), Iterative refinement of least squares solutions I, *B.I.T.* **7**, 257-278. *117*

Björck, A. (1968). Iterative refinement of least squares solutions II, *B.I.T.* **8**, 8-30. *117 131*

Björck, A. and Golub, G. H. (1967). Iterative refinement of least squares solution by Householder transformations, *B.I.T.* **7**, 322-337. *131*

Bowdler, H., Martin, R. S., Reinsch, C. and Wilkinson, J. H. (1968). The QL and QR algorithms for symmetric matrices. *Numer. Math.* **11**, 293-306; *HAC(II)* 227-240. *113*

Businger, P. A. (1965). Algorithms 253 and 254. Eigenvalues and eigenvectors of a symmetric matrix. *Comm. A.C.M.* **8**, 218-219. (Also the certification by J. H. Welsch, *ibid* **10**, (1967). 376.) *113*

Businger, P. A. (1970). Updating a singular value decomposition. *B.I.T.* **10**, 376-385. *123*

Businger, P. A. and Golub, G. H. (1965). Linear least squares solutions by Householder transformation. *Numer. Math.* **7**, 269-276; *HAC(II),* 111-118. *130*

Businger, P. A. and Golub, G. H. (1969). Algorithm 358: Singular value decomposition of a complex matrix. *Comm. A.C.M.* **12**, 564-565. *112*

Chambers, J. M. (1971). Regression Updating. *J. Amer. Statist. Assoc.* **66**, 744-748. *123 131*

Chambers, J. M. (1972). Stabilizing linear regression against observational error in independent variates. *Bell Laboratories Technical Memorandum.* Bell Laboratories, Murray Hill NJ. *117*

Chambers, J. M. (1974). Linear regression computations: some numerical and statistical aspects. *Proc. Int. Stat. Inst.* **39**, Book 4, 246-254. *120*

Chambers, J. M. (1975). Updating Linear models for the addition and deletion of observations. In *Experimental Design and Linear Models,* J. N. Srivastava (ed.), North Holland Press. *122 123*

Chambers, J. M. (1975a), Numerical Methods for the Analysis of Variance. Bell Laboratories, Murray Hill, NJ. *129*

Claringbold (1969). Algorithm AS22: The interaction algorithm. *Appl. Statist.* **18**, 75-79. *127*

Clayton, D. G. (1971). Algorithm AS46: Gram-Schmidt orthogonalization. *Appl. Statist.* **20**, 335-338. *130*

Daniel, J. W., Gragg, W. B., Kaufman, L., and Stewart, G. W. (1976). Reorthogonalization and stable algorithms for updating the Gram-Schmidt QR factorization. *Math. Comp.*, **30**, 772-795. *105 107 116 120 123 130*

Dempster, A. P. (1969). *Elements of Continuous Multivariate Analysis.* Addison-Wesley, Reading, MA. *125*

Fowlkes, E. B. (1969). Some operators for ANOVA calcuations. *Technometrics* **11**, 511-526. *129*

Gentleman, W. M. (1973). Least squares computations by Givens transformations without square roots. *J. Inst. Maths Applics.* **12**, 329-336. *120 123*

Gentleman, W. M. (1974). Basic procedures for large, sparse or weighted least-squares, *Appl. Statist.*, **23**, 448-454. *116 123 131*

Gnanadesikan, R. (1977). *Methods of Statistical Data Analysis of Multivariate Observations.* Wiley, New York. *101 125*

Golub, G. H. and Kahan, W. (1966). Calculating the singular values and pseudo-inverse of a matrix. *SIAM J. Numer. Anal.* **2**, 202-224. *131*

Golub, G. H., Klema, V., and Stewart, G. W. (1977). Rank degeneracy and the least-squares problem. Submitted to *J. Amer. Statist. Assoc.* *117*

Golub, G. H. and Reinsch, C. (1970). Handbook series linear algebra: Singular value decomposition and least squares solutions. *Numer. Math.* **14**, 403-420. *HAC(II),* 134-151. *112 131*

Golub, G. H. and Styan, G. P. H. (1973). Numerical computations for univariate linear models. *J. Statist. Comput. Simul.* **2**, 253-274. *114 123*

Gower, J. C. (1969). Algorithm AS18: Evaluation of marginal means; Algorithm AS19: Analysis of variance for a factorial table. *Appl. Statist.* **18**, 197-202. *129*

Gower, J. C. (1969a), Algorithm AS23: Calculation of effects. *Appl. Statist.* **18**, 79-82. *127*

Hampel, F. R. (1973). Robust estimation: A condensed partial survey. *Z. Warsch.* **27**, 87-104. *124*

Hemmerle, W. J. (1974). Non-orthogonal analysis of variance using iterative improvement and balanced residuals. *J. Amer. Statist.*

Assoc. **69**, 772-778. *129*

Hoaglin, D. C. and Welsch, R. E. (1977). The hat matrix in regression and anova. Submitted to *Amer. Statistician.* *110*

Howell, J. R. (1969). Algorithm 359: Factorial analysis of variance. *Comm. A.C.M.* **12**, 631-632. *127*

Huber, P. J. (1973). Robust regression: asymptotics, conjectures and Monte Carlo. *Ann. Stat.* **1**, 799-821. *124*

James, A. T. and Wilkinson, G. N. (1971). Factorization of the residual operator and canonical decomposition of nonorthogonal factors in the analysis of variance. *Biometrika* **58**, 279-294. *129*

Kaufman, L. (1975). The LZ algorithm to solve the generalized eigenvalue problem for complex matrices. *Trans. on Math. Software* **1**, 271-281. *113*

Lawson, C. L. and Hanson, R. J. (1974). *Linear Least-Squares Solutions.* Prentice-Hall, Englewood Cliffs, NJ. *101 112 116 120 123 130 131*

Lide, D. R., Jr. and Paul, M. A. (1974). (eds.) *Critical Evaluation of Chemical and Physical Structural Information.* National Academy of Sciences, Washington. *124*

Longley, J. W. (1967). An appraisal of least-squares programs from the point of view of the user. *J. Amer. Statist. Assoc.* **62**, 819-841. *117 132*

Martin, R. S. and Wilkinson, J. H. (1968). The implicit QL algorithm. *Numer. Math.* **12**, 377-383; *HAC(II)* 241-248. *113*

Moler, C. B. and Stewart, G. W. (1973). An algorithm for generalized matrix eigenvalue problems. *SIAM J. Numer. Anal.* **10**, 241-256. *101 113 114*

Oliver, I. (1968). Algorithm 330: Factorial analysis of variance. *Comm. A.C.M.* **11**, 431-432. *127*

Preece, D. A. (1971). Iterative procedures for missing values in experiments. *Technometrics* **13**, 743-753. *129*

Rao, C. R. (1973). *Linear Statistical Inference and its Applications.* Second Edition. Wiley, New York. *101 102 125*

Rubin, D. B. (1972). A non-iterative algorithm for least squares estimation of missing values in any analysis of variance design. *Appl. Statist.* **21**, 136-141. *129*

Scheffe, H. (1959). *The Analysis of Variance,* Wiley, New York. *127*

Stewart, G. W. (1973). *Introduction to Matrix Computations.* Academic, New York. *101 114*

Wilkinson, G. N. (1970). A general recursive procedure for analysis of variance. *Biometrika* **57**, 19-46. *129*

Wilkinson, G. N. and Rogers, C. E. (1973). Symbolic description of factorial models for analysis of variance. *Appl. Statist.* **22**, 392-399. *129*

Wilkinson, J. H. (1965). *The Algebraic Eigenvalue Problem.* Clarendon Press, Oxford. *120*

Wilkinson, J. H. and Reinsch, C. (eds.) (1971). *Handbook for Automatic Computation. Volume II. Linear Algebra.* Springer Verlag, Berlin. (Papers in this volume are cross-referenced as *HAC(II)* followed by page numbers.)

Yates, F. (1937). The design and analysis of factorial experiments. *Imp. Bur. Soil. Sci. Tech. Comm.* 35. *127*

Chapter 6

Bartels, R. H., and Golub, G. H. (1969). Algorithm 350: Simplex procedure employing LU decomposition. *Comm. A. C. M.,* **12**, 275-277. *158*

Beale, E. M. L. (1967). Decomposition and partitioning methods for nonlinear programming. *Numerical Programming,* J. Abadie (ed.), Wiley, New York, 135-205. *159*

Bickel, P. J. (1975). One-step Huber estimates in the linear model. *J. Amer. Statist. Assoc.* **70**, 428-434. *136*

Björck, A. (1968). Iterative refinement of linear least - squares solutions *B.I.T.* **8**, 8-30. *159*

Björck, A. and Golub, G. H. (1967). Iterative refinement of least squares solution by Householder transformations, *B.I.T.* 7, 322-337. *159*

Box, M. J. (1965). A new method of constrained optimization and a comparison with other methods, *Computer J.* **8**, 42-52. *158*

Brent, R. P. (1973). *Algorithms for Minimization without Derivatives* Prenctice Hall, Englewood Cliffs, NJ. *141 142*

Broyden, C. G. (1967). Quasi-Newton methods and their application to function minimization, *Math. Comp.* **21**, 368-381. *140*

Broyden, C. G., Dennis, J. E. Jr., and More, J. J. (1973). On the local and superlinear convergence of quasi-Newton methods. *J. Inst. Maths Appl.* **12**, 223-244. *145 153*

Chambers, J. M. and Ertel J. E. (1974). Remark AS R11 *Appl. Statist.* **23**, 250-251. *142*

Crowder, M. J. (1976). Maximum likelihood estimation for dependent observations. *J. Roy. Statist. Soc. (B),* **38**, 45-53. *148*

Cohen, A. I. (1972). Rate of convergence of several conjugate gradient algorithms *SIAM J. Numer. Anal.* **9**, 248-259. *144*

Conn, A. R. and Pietrzykowski, T. (1977). A penalty function method converging directly to a constrained optimum. *SIAM J. Numer. Anal.* **14**, 348-375. *157*

Dantzig, G. B. (1963). *Linear Programming and Extensions,* Princeton University Press, Princeton, NJ. *159*

Davidon, W. C. (1959). Variable metric method for minimization, *Argonne Nat. Lab. Report 5990 (rev).* *138*

Davidon, W. C. (1975). Optimally conditioned optimization optimization algorithms without line searches. *Math. Prog.* **9**, 1-30. *140 141*

Dixon, L. C. W. (1972). Variable metric algorithms: Necessary and sufficient conditions for identical behavior of nonequadratic functions. *J. Opt. Th. Applic.* **10**, 34-40. *140 144 145*

Dixon, L.C.W. (1972a). The choice of step length, a crucial factor in the performance of variable metric algorithms. *Numerical Methods for Nonlinear Optimization,* F. A. Lootsma (ed.), Academic, London, 149-170. *140*

Ertel, J. E. (1975). *Asymptotic Properties of Estimates Obtained by Nonlinear Optimization or by Solving Nonlinear Equations.* Doctoral Thesis, Princeton University, Department of Statistics. *148*

Fiacco, A. V. and McCormick, G.P. (1968). *Sequential Unconstrained Minimization Techniques for Nonlinear Programming,* Wiley, New York. *157*

Fisher, R. A. (1922). On the mathematical foundations of theoretical statistics, *Phil. Trans. Roy. Soc. London (A)* **222**. 309-368. *147*

Fleming, W. H. (1965). *Functions of Several Variables,* Addison-Wesley, Reading, MA. *154 156*

Fletcher, R. (1966). Certification of algorithm 251: Function minimization, *Comm. A.C.M.* **9**, 686. *139*

Fletcher, R. (1970). A new approach to variable metric algorithms, *Computer J.* **13**, 317-322. *140 141 145*

Fletcher, R. (1970a). A class of methods for nonlinear programming with termination and convergence properties. In Abadie (ed.) *Integer and Non-linear Programming,* North Holland Press. *158*

Fletcher, R. (1971). A modified Marquardt subroutine for nonlinear least squares. *Tech. Report AERE R6799* H.M. Stationery Office, London

Fletcher, R. (1972). A class of methods for nonlinear programming III. Rates of convergence. *Numerical Methods for Nonlinear Optimization,* F. A. Lootsma (ed.), Academic, London, 371-382. *158*

Fletcher, R. (1973). An exact penalty function for nonlinear programming with inequalities. *Math. Prog.* **5**, 129-150. *157 158*

Fletcher, R. and Lill, S.A. (1970). A class of methods for nonlinear programming II. Computational experience. *Symposium on Nonlinear Programming* J. B. Rosen (ed.), Academic, 67-92. *139 158*

Fletcher, R. and Powell, M.J.D. (1963). A rapidly convergent descent method for minimization, *Computer J.* **6**, 163-168. *138 144*

Fletcher, R. and Reeves, C. M. (1964). Function minimization by conjugate gradients, *Computer J.* **7**, 149-154. *142*

Garcia-Palomares, U. M. and Mangasarian, O. L. (1976). Superlinearly convergent quasi-Newton algorithms for nonlineary constrained optimization problems. *Mathematical Programming,* **11**, 1-13. *157*

Gill, P. E. and Murray, W. (1970). A numerically stable form of the simplex algorithm. *Report Math 87,* National Physical Laboratory, England. *158*

Gill, P. E. and Murray, W. (1972). Quasi-Newton methods for unconstrained optimization. *J. Inst. Maths Applics,* **9**, 91-108. *140*

Goldfeld, S. M., Quandt, R. E. and Trotter, H. F. (1966). Maximization by quadratic hill-climbing, *Econometrica* **34**, 541-551. *138*

Golub G. H. and Pereyra, V. (1973). The differentiation of pseudoinverses and nonlinear least squares problems whose variables separate. *SIAM J. Numer. Anal.* **10**, 410-423. *151 155*

Greenstadt, J. L. (1966). A ricocheting gradient method for nonlinear optimization, *SIAM J. Appl. Math.* **14**, 429-445. *158*

Haarhoff, P. C. and Buys, J. D. (1970). A new method for the optimization of a nonlinear function subject to nonlinear constraints, *Computer J.* **13**, 178-184. *158*

Hamilton, P. A. and Boothroyd, J. (1969). Certification of algorithm 251: Function minimization, *Comm. A.C.M.* **12**, 512. *139*

Hartley, H. O. (1961). The modified Gauss-Newton method for the fitting of non-linear regression functions by least squares *Technometrics* **3**, 269-280. *150*

Himmelblau, D.M. (1972). *Applied Nonlinear Programming.* McGraw Hill, New York. *159*

Jennrich, R. I. (1969). Asymptotic properties of nonlinear least squares estimates. *Ann. Math. Statist.* **41**, 956-969. *148*

Kaufman, L. (1975). A variable projection method for solving separable nonlinear least-squares problems. *B. I. T.,* **15**, 49-57. *151*

Kruskal, J. B. and Carroll, J. D. (1969), Geometric models and badness-of-fit functions, *Multivariate Analysis, II* (P.R. Krishnaiah, ed.), Academic, New York. *135*

Lill, A. S. (1970). Algorithm 46. A modified Davidon method for finding the minimum of a function, using difference approximation for derivatives. *Computer J.* **13**, 111-113 (also *ibid* **14**, (1971). p. 106, and p. 214). *141*

Luenberger, D. G. (1974). *Introduction to Linear and Nonlinear Programming.* Addison-Wesley, Reading, MA. *159*

Mangasarian, O. (1969). *Nonlinear Programming.* McGraw Hill, New York *156*

Marquardt, D. W. (1963). An algorithm for least squares estimation of nonlinear parameters, *SIAM J. Appl. Math.* **11**, 431-441. *151*

Marquardt, D. W. (1970). Generalized inverses, ridge regression, biased linear estimation and nonlinear estimation, *Technometrics* **12**, 591-612. *151*

Murtagh, B. A. and Sargent, R. W. H. (1969). A constrained minimization method with quadratic convergence. *Optimization.* (R. Fletcher, ed.) Academic, London, 215-246. *158*

Nelder, J. A. and Mead, R. (1965). A simplex method for function minimization. *Computer J.* **7**, 308-313. *142 159*

Nelder, J. A. and Wedderburn, R. W. M. (1972). Generalized linear models *J. R. Statist. Soc. (A)* **135**, 370-384. *153*

O'Neill, R. (1971). Algorithm AS47: Function minimization using a simplex procedure. *Appl. Statist.* **20**, 338-345. *142*

Ortega, J. M. and Rheinboldt, W. C. (1970). *Iterative Solution of Nonlinear Equations in Several Variables* Academic, New York. *153 155*

Polak, E. (1971). *Computational Methods in Optimization.* Academic, New York. *144 145*

Polak, E. and Ribiere, G. (1969). Note sur la convergence de methodes de direction conjugees. *Rev. Franc. Inf. Rech. Oper.* **16**, 35-43. *143 144*

Powell, M. J. D. (1964). An efficient method for finding the minimum of a function of several variables without calculating derivatives, *Computer J.* **7**, 155-162. *141 145*

Powell, M. J. D. (1969). A method for nonlinear constraints in minimization problems. *Optimization.* (R. Fletcher, ed.) Academic, London, 283-298. *158*

Powell, M. J. D. (1970). A survey of numerical methods for unconstrained optimization, *SIAM Rev.* **12**, 79-97.

Powell, M. J. D. (1970a). A new algorithm for unconstrained optimization. *Nonlinear Programming.* (J. B. Rosen et al, eds) Academic, New York. *145*

Powell, M. J. D. (1970b). A fortran subroutine for unconstrained minimization, requiring first derivatives of the objective function, Report A.E.R.E.-R6469 (available from Her Majesty's Stationery Office, London). *145*

Powell, M. J. D. (1971). On the convergence of the variable metric algorithm. *J. Inst. Maths Appl.* **7**, 21-36. *145*

Powell, M. J. D. (1972). Quadratic termination properties of minimization algorithms I and II *J. Inst. Maths Appl.* **10**, 333-342, 343-357. *145*

Powell, M. J. D. (1972a). Some properties of the variable metric algorithms. *Numerical Methods for Nonlinear Optimization,* F. A. Lootsma (ed.), Academic, London, 1-17. *145 144 145*

Powell, M. J. D. (1976). Some convergence properties of the conjugate gradient method. *Math. Programming,* **11**, 42-29. *144*

Richardson, J. A. and Kuester, J. L. (1973). Algorithm 454, The complex method for contrained optimization *Comm.A.C.M.* **16**, 487-489. *158*

Rusin, M. H. (1971). A revised simplex method for quadratic programming *SIAM J. Appl. Math.* **20**, 143-160. *159*

Shanno, D. F. (1970). Conditioning of quasi-Newton methods for function minimization. *Math. Comp.,* **24**, 647-656. *140*

Shanno, D. F. and Phua, K. H. (1976). Algorithm 500: Minimization of Unconstrained Multivariate Functions. *Trans. on Math. Software* **2**, 87-94. *141*

Spendley, W., Hext, G. R. and Himsworth, F. R. (1962). Sequential applicationss of simplex designs in optimization and evolutionary operation, *Technometrics* **4**, 441-461. *158*

Spivey, W. A. and Thrall, R. M. (1970). *Linear Programming.* Prentice-Hall, Englewood Cliffs, NJ. *159*

Springer, B. G. F. (1971). Numerical optimization in the presence of random variability. The multi-factor case. Unpublished paper, Faculty of Agriculture, University of West Indies, Trinidad. See also his Ph. D. thesis, Dep't of Mathematics, Imperial College (1968). *142*

Wells, M. (1965). Algorithm 251: Function minimization. *Comm. A.C.M.* **8**, 169. See also Fletcher (1966). and Hamilton and Boothroyd (1969).. *139*

Wold, H. (1966). Nonlinear estimation by iterative least squares procedures. *Festschrift for Jerzy Neyman.* F. N. David (ed.) Wiley, New York 411-444. *153*

Wold, H. (1973). Nonlinear iterative partial least squares (NIPALS) modelling. *Multivariate Analysis (III)* (P. Krishnaiah, ed.) Academic, New York 383-407. *153*

Zoutendijk, M. (1960). *Methods of Feasible Direction.* Elsevier, Amsterdam. *158*

Chapter 7

Ahrens, J. H. and Dieter, U. (1972). Computer methods for sampling from the exponential and normal distributions. *Comm. A. C. M.,* **15**, 873-882. *180*

Ahrens, J. H. and Dieter, U. (1974). Computer methods for sampling from gamma, beta, Poisson and binomial distributions. *Computing,* **12**, 223-246. *183*

Ahrens, J. H. and Dieter, U. (1974a) Acceptance-rejection techniques for sampling from the gamma and beta distributions. Technical report, AD-782478, Stanford University. (available from National Technical Information Service).

Andrews, D. F., Bickel, P. J., Hampel, F. R., Huber, P. J., Rogers, W. H. and Tukey, J. W. (1972). *Robust Estimates of Location: Survey and Advances.* Princeton University Press. *173*

Brent, R. P. (1974). Algorithm 488: A Gaussian pseudo-random number generator. *Comm. A. C. M.* **17**, 704-706.

Chambers, J. M. (1970). Computers in statistical research: Simulation and computer-aided mathematics. *Technometrics* **12**, 1-15. *186*

Chambers, J. M., Mallows, C. L. and Stuck, B. W. (1976). A method for simulating stable random variables. *J. Amer. Statist. Assoc.* **71**, 340-344. *182*

Chambers, R. P. (1967). Random number generation. *IEEE Spectrum* **4**, part 1, no. 2, 48-56. *169*

Coveyou, R. R. and Macpherson, R. D. (1967). Fourier analysis of uniform random number generators. *J. A. C. M.* **14**, 100-119. *167 174*

Feller, W. (1971). *An Introduction to Probability Theory and its Applications.* John Wiley and Sons, New York. *182*

Golomb, S. W. (1967). *Shift Register Sequences.* Holden-Day, San Francisco. *171*

Gross, A. M. (1973). A Monte Carlo swindle for estimators of location. *Appl. Statist.* **22**, 347-353. *189*

Gross, A. M. (1976). Portable random number generation. *Unpublished Memorandum,* Bell Laboratories, Murray Hill, NJ. (See also Problem 2). *174 191*

Halton, J. H. (1970). A retrospective and Prospective survey of the Monte Carlo method. *SIAM Review* **12**, 1-63. *189*

Hammersley, J. M. and Handscomb, D. C. (1964). *Monte Carlo Methods* Methuen, London. *190*

Hastings, W. K. (1970). Monte Carlo sampling methods using Markov chains and their applications. *Biometrika* **57**, 97-109. *190*

Jansson, B. (1966). *Randon Number Generators* Victor Pettersons Bokindustri Akt., Stockholm. *163 165 166 170*

Jenkins, G. M. and Watts, D. G. (1968). *Spectral Analysis and its Applications.* Holden-Day, San Francisco. *167*

Kinderman, A. J. and Ramage, J. G. (1976). Computer generation of normal random variables. *J. Amer. Statist. Assoc.,* **71**, 893-896. *180*

Knuth, Donald E. (1969). *The Art of Computer Programming* **2**, (Semi-numerical Algorithms) Addison Wesley, Reading MA. *162 165 170 172 174 178 183*

Lewis, P. A. W., Goodman, A. S. and Miller, J. M. (1969). A pseudo-random number generator for the System/360. *IBM Systems J.* **8**, 136-146. *196*

Lewis, T. G. and Payne, W. H. (1973). Generalized feedback shift register pseudorandom number algorithm. *J. A. C. M.* **20**, 456-468. *172*

MacLaren, M. D. and Marsaglia, G. (1965). Uniform random number generators. *J. A. C. M.,* **12**, 83-89. *174*

Maclaren, M. D., Marsaglia G. and Bray, T. A. (1964). A fast procedure for generating exponential random variables. *Comm. A. C. M.* **7**, 298-300. *178 180*

Marsaglia, G.(1972). The structure of linear congruential sequences. In *Applications of Number Theory to Numerical Analysis.* Academic, New York. *164 165 168 169 173 174*

Marsaglia, G. (1972a). Choosing a point from the surface of a sphere. *Annals Math. Stat.* **43**, 645-646. *192*

Marsaglia, G., Ananthanarayanan, K. and Paul, N. (1973). The McGill random number package "Super-Duper". Available from School of Computer Science, McGill University, Montreal, CANADA.

Marsaglia, G., Maclaren, M. D. and Bray, T. A. (1964). A fast procedure for generating normal random variables. *Comm. A. C. M.*

7, 4-10. *178*

Minsky, Marvin (1967). *Computation: Finite and Infinite Machines* Prentice-Hall, Englewood Cliffs, NJ *162*

Murry, H. F. (1970). A general approach for generating natural random variables. *I.E.E.E. Trans. on Comp.* **19**, 1210-1213. *170*

Ramberg, J. S. and Schmeiser, B. W. (1972). An approximate method for generating symmetric random variables. *Comm. A. C. M.* **15**, 987-990. *179*

RAND Corporation (1955). *One Million Random Digits and 100 000 Normal Deviates.* Free Press, New York. *171*

Relles, D. A. (1970). Variance reduction techniques for Monte Carlo sampling from Student distributions. *Technometrics* **12**, 499-515. *189*

Snow, R. H. (1968). Algorithm 342: Generation of random numbers satisfying the Poisson distribution. *Comm. A. C. M.,* **11**, 281. *183*

Tausworthe, R. C. (1965). Random numbers generated by linear recurrence modulo two. *Math. Comp.* **19**, 201-209. *172 173*

Tootill, J. P. R., Robinson, W. D. and Adams, A. G. (1971). The runs up-and-down performance of Tausworthe pseud0-random number generators. *J. A. C. M.* **18**, 381-399. *173 174*

von Mises, R. (1957). *Probability, Statistics and Truth* (2nd English edition). Macmillan, New York. *161*

Chapter 8

Becker, R. A. and Chambers, J. M. (1976). On structure and portability in graphics for data analysis. *Proc. Ninth Interface Symposium on Computer Science and Statistics.* Prindle, Weber and Schmidt, Boston. *194*

Bell Laboratories (1977). *The GR-Z system of graphical subroutines for data analysis.* (See Appendix on Algorithms). *194 221 225*

Chambers, J. M. (1973). Color Contour Plots: Interactive Study of Function Surfaces. *Unpublished Bell Laboratories Memorandum. 214 218*

Chambers, J. M.(1975). Structured computational graphics for data analysis. *Proc. Int. Statist. Inst.* **40**, 467-486. *194*

Crane, C. M. (1972). Contour plotting for functions specified at nodal points of an irregular mesh based on an arbitrary two-parameter co-ordinate system. *Computer J.* **15**, 382-384. *217 226*

Doane, D. P. (1976). Aesthetic frequency classifications. *Amer. Statistician* **30**, 181-183. *222 226*

Graham, N. Y. (1972). Perspective drawing of surfaces with hidden line elimination. *Bell Sys. Tech. J.* **51**, 843-861. *212*

Knowlton, K. and Harmon, L. (1972). Computer-produced grey scales. *Computer Graphics and Image Processing.* **1**, 1-20. *199 214*

Lewart, C. R. (1973). Algorithm 463: Algorithms SCALE1, SCALE2, and SCALE3 for determination of scales on computer generated plots. *Comm. A. C. M.* **16**, 639-640. *219 226*

Marlow, S. and Powell, M. J. D. (1976). A FORTRAN subroutine for plotting the part of a conic that is inside a given triangle. *Report A.E.R.E.-R.8336,* Atomic Energy Research Establishment, Harwell. *216*

McConalogue, D. J. (1970). A quasi-intrinsic scheme for passing a smooth curve through a discrete set of points. *Computer J.* **13**, 392-396. *209*

McConalogue, D. J. (1971). An automatic French-Curve procedure for use with an incremental plotter. *Computer J.* **14**, 207-209. *209*

Newman, W. M. and Sproull, R. F. (1973). *Principles of Interactive Computer Graphics.* McGraw-Hill. *211*

Sutherland, I. E., Sproull, R. F. and Schumaker, R. A. (1974). A characterization of ten hidden-surface algorithms. *Computing Surveys* **6**, 1-55. *211 212*

Tukey, J. W. (1977). *Exploratory Data Analysis.* Addison-Wesley, Reading MA. *223*

Williamson, H. (1972). Algorithm 420: Hidden-line plotting program. *Comm. A. C. M.* **15**, 100-103. *212 226*

Wilk, M. B. and Gnanadesikan, R. (1968). Probability plotting methods for the analysis of data. *Biometrika,* **55**, 1-17. *223 225*

Wilson, L. O. (1974). The shielding of a plane wave by a cylindrical array of infinitely long thein wires. *Trans. I.E.E.E. on Propogation and Transmission* (to appear). *217*

Wright, T. (1974). Algorithm 475: Visible surface plotting program. *Comm. A. C. M.* **17**, 152-155. (See also the Remarks on the above, *Comm. A.C.M.* **18**, (1975). 276-277, and *Trans. Math. Software.,* **2**, (1976)., 109-110.) *212 226*

APPENDIX

Available Algorithms

The following tables are a selected list of algorithms related to the methods discussed in the book. By *algorithm* we mean some definition of a computation which can either be run on a computer directly (i. e., a program) or else can be converted into a program by following a clear and complete set of instructions. The selection is based primarily on the quality of the algorithm, as judged by a combination of published evaluations, advice and personal experience. A degree of subjectivity remains, and the selection is not claimed to represent a thorough evaluation of all competing algorithms. Preference has been given to published algorithms (which, one hopes, have been tested), and also to algorithms which are easy to implement and use. Where algorithms are available in machine-readable form, a brief explanation of how to get them is included. The key words (BTL, EISPACK, etc) in the last column of the tables are explained, beginning on page 253, under **How to Get Them**. References will be found in the References under the corresponding Chapter.

Chapter 2: Programming.

Algorithm	Reference	Language	How to get it
Profiling	Sande (1975) OE02A	FORTRAN FORTRAN	Note 1 HARWELL
Portable-FORTRAN Verifier	Ryder (1974)	FORTRAN	BTL
Structured FOR-TRAN	Kernighan and Plauger (1976)	RATFOR	Note 2

Notes.

1. For information on the PROFILER program, write to:

   Gordon Sande
   General Survey Systems
   Statistics Canada
   Tunneys Pasture
   Ottawa, CANADA

2. In addition to the specific algorithms above, a number of utilities are helpful to writing good programs. Where possible, these should be provided by the operating system. Otherwise, the user may supply at least partial replacements. Kernighan and Plauger (1976) discuss the writing of a number of such programs, and supply actual examples. Machine-readable copies of the programs are available from their publisher.

Chapter 3: Data Management and Manipulation.

Algorithm	Reference	Language	How to get it
Data Management			
Stack Storage	Fox et al (1975)	FORTRAN	Listing, PORT
List Management	Bray (1974)	FORTRAN	Listing
Storage Management	Knuth (1968, 2.5)	Description	
Sorting			
Internal Sort	Singleton (1969), modified in Loesser (1976)	ALGOL60, FORTRAN	Listing
Partial Sort	Chambers (1971)	FORTRAN	Listing
Table Look-up	Brent (1973)	FORTRAN	Listing

Chapter 4: Numerical Methods.

Algorithm	Reference	Language	How to get it
Vector length	Blue (1977)	FORTRAN	PORT
Rational Approximation			
Remes' Algorithm	Cody et al (1969) Werner et al (1969)	ALGOL60	Listing
Differential Correction		FORTRAN	PORT
Splines			
B-splines		FORTRAN	PORT NPL
Numerical Integration			
Clenshaw-Curtis Method	Gentleman (1972a)	FORTRAN	Listing
Adaptive Romberg	Blue (1975)	FORTRAN	Listing, PORT
Spectral Analysis			
Fourier Transform	Singleton (1969)	FORTRAN	Listing
Auxiliary Calculations	Bloomfield (1976)	FORTRAN	Listing
		FORTRAN	IMSL

Chapter 5: Linear Models.

Algorithm	Reference	Language	How to get it
Orthogonal Decompositions			
Gram-Schmidt	Daniel et al (1976)	ALGOL60	Listing
Householder	Lawson and Hanson (1974), Businger and Golub (1965)	FORTRAN, ALGOL60	Listing
Givens Method	Gentleman (1974)	ALGOL60	Listing
	Chambers (1971)	FORTRAN	
Singular-Value Decomposition	Golub and Rheinsch (1970)	ALGOL60	Listing
		FORTRAN	ROSEPACK, LINPACK
Eigenvalue Decomposition (various)	EISPACK	FORTRAN	EISPACK
	Wilkinson and Rheinsch (1971)	ALGOL60	Listing
L_1 Regression	Barrodale and Roberts (1974)	FORTRAN	Listing, ACM
	Bartels and Conn (1978)	FORTRAN	ROSEPACK, ACM
Analysis of Variance			
Yates Method	Claringbold (1969)	FORTRAN	Listing
	Howell (1969)	FORTRAN	
	Oliver (1968)	ALGOL60	
Other Methods	Gower (1969)	ALGOL60	Listing
		FORTRAN	IMSL

Chapter 6: Nonlinear Models.

Algorithm	Reference	Language	How to get it
Optimization			
Quasi-Newton	Hua and Shanno (1976)	FORTRAN	ACM
with factorization	HARWELL	FORTRAN	HARWELL
Search and Simplex	O'Neill (1971)	FORTRAN	Listing
	Brent (1973)	ALGOL-W	
Nonlinear Least-Squares	Fletcher (1971)	FORTRAN	Listing

Chapter 7: Simulation of Random Processes.

Algorithm	Reference	Language	How to get it
Pseudorandom Uniforms			
Exclusive-or	Marsaglia et al. (1973),	IBM 360 Assembler, FORTRAN	
	Gross (1976)	FORTRAN	Listing
Replacement	Knuth (1969, p. 30)	Description	
Shuffle	Andrews et al (1972)	PL 360	Listing.
Normal Distribution			
Box-Wedge-Tail	Marsaglia et al (1964)	Description	
Mixture	Brent (1974)	FORTRAN	Listing, ACM
	Kinderman and Ramage (1976)	Description	
Exponential	Maclaren et al (1964)	Description	
Chi-Square, Gamma, F, Beta and Binomial	Ahrens and Dieter (1974; 1974a)	Description	
Stable Family	Chambers et al (1976)	FORTRAN	Listing
Poisson	Knuth (1969, 117-118), Ahrens and Dieter (1974)	Description	
Multivariate Normal	Chambers (1970)	Description	

Chapter 8: Computational Graphics.

Algorithm	Reference	Language	How to get it
Plotting Curves (see also Splines)	Mconalogue (1971)	ALGOL60	Listing
Surfaces with hidden lines removed	Williamson (1972), Wright (1974)	FORTRAN	Listing
Contours	Crane (1972)	ALGOL60	Listing
Choice of Scale	Lewart (1973)	FORTRAN	Listing
Histogram scale	Doane (1976)	Description	

Note.

Graphical computations benefit particularly from using a co-ordinated set of algorithms. A number of such are available, for example Bell Laboratories (1977).

How to get them.

ACM The Association for Computing Machinery Collected Algorithm Series has been published since 1960, first in the *Communications,* then in *Transactions on Mathematical Software.* Beginning in 1974, copies of the algorithms were distributed in machine-readable form. From 1975, the distribution policy was to publish only the preliminary comments (calling sequence, etc) without the program listing, for longer algorithms and packages. Listings may be obtained for all algorithms by subscribing to the Collected Algorithms service, or for individual algorithms be request. Card decks or magnetic tape copies of the source for individual algorithms may be obtained on request. Fees are charged, designed to recover costs. Orders may be sent to:

ACM Algorithms Distribution Service
c/o IMSL, Inc.
Sixth Floor, GNB Building
7500 Bellaire Boulevard
Houston, TX 77036
USA

ARGONNE The Argonne Code Center distributes a number of packages of numerical software, including packages for eigenanalysis, linear equations and function approximation. Originally, the programs were distributed royalty-free. Recently, some changes in policy have occured; potential users should enquire as to current charges. Representatives of installations interested in obtaining one or more of the packages should write for information to:

Argonne Code Center
Argonne National Laboratory
9700 S. Cass Avenue
Argonne, IL 60439
USA

Note that Argonne packages are provided for specific machines. The number of different machines supported and the degree of machine dependence vary with the package (contrast PORT below).

BTL Some Bell Laboratories software systems are available on a license basis. A nonprofit educational institution may obtain a royalty-free license for a specific system, to be used for educational purposes. There is a small service charge to help defray the cost of distribution. Inquiries for an educational license should be directed to:

Bell Laboratories
Computing Information Service
600 Mountain Avenue
Murray Hill, New Jersey 07974
USA

For commercial and governmental organizations, and for educational institutions desiring to use the system for commercial purposes, a royalty is charged. Inquiries should be directed to:

Western Electric Co.
Patent Licensing
P.O. Box 20046
Greensboro NC 27420
USA

EISPACK This is a package of eigenvalue and related algorithms, distributed by Argonne Laboratories (see ARGONNE above). The package is distributed in versions for specific machines. Included are FORTRAN versions of many of the ALGOL60 algorithms from *Numerische Mathematik* collected in Wilkinson and Rheinsch (1971). A general guide to the routines is the book *Matrix Eigensystem Routines - EISPACK Guide* by B. T. Smith et al., Second Edition, 1976, Springer-Verlag, New York.

IMSL This is a library of mathematical and statistical utilities and numercial algorithms. The choice and organization of procedures are more closely related to conventional statistical formulation of problems than in the case of many of the other libraries. The library is available for an annual royalty. For further information, write:

International Mathematical and
Statistical Libraries, Inc.
Sixth Floor, GNB Building
7500 Bellaire Boulevard
Houston, TX 77036
USA

LINPACK This is a package of portable FORTRAN routines for the solution of linear equations, linear least squares and related computations. As this is written, the package is scheduled to appear in 1978. Details of distribution are not entirely determined. For further information write to:

LINPACK Project
Argonne National Laboratory
9700 S. Cass Avenue
Argonne, IL 60439
USA

HARWELL The U. K. Atomic Energy Research Establishment maintains and distributes a library of numerical and other algorithms. The source code is written for an IBM 70 system, although some of the algorithms can be moved to other machines which have compatible FORTRAN compilers without great effort. The presence at Harwell of eminent workers in nonlinear model-fitting has in the past provided a particularly rich set of algorithms in this field. Charges for copies of source code are made on the basis of recovering costs. Those interested should write to:

Mr. S. Marlow
Building 8.9
Harwell, Didcot
OX11 ORA
ENGLAND

NPL The National Physical Laboratory maintains a library of numerical algorithms. Of particular interest for data analysis are a number of routines for approximation and nonlinear fitting. Algorithms are available in ALGOL60 or FORTRAN. While some of the older algorithms are in ALGOL60 only, the current policy is to provide all future routines in both languages. Portability of programs is encouraged by conventions to be followed by programmers, but it is not required for an algorithm to be included in the library. For further information, write:

Division of Numerical Analysis
and Computing
National Physical Laboratory
Teddington, Middlesex
ENGLAND TW11 0LW

PORT is a library of FORTRAN subprograms for numerical computation, developed at Bell Laboratories. The library is portable across differen: machines, and includes dynamic storage allocation, error checking routines and some other utilities. Portability is attained through adherence to a portable subset of FORTRAN (Ryder, 1974) and through the use of functions which return the values of machine-dependent constants. Thus, redefining the constants effectively implements the code on a new machine. For access to PORT, see BTL above.

ROSEPACK This is a subroutine package with routines relating to several areas of statistical analysis, notably robust regression and related topics. The routines are produced in FORTRAN for specific machines, based on source code in extended FORTRAN. For further information, write

ROSEPACK Library
National Bureau of Econ. Res.
Computer Research Center
575 Technology Square
Cambridge, MA 02139
USA

INDEX

Applied Probability and Statistics (Continued)

HAHN and SHAPIRO · Statistical Models in Engineering
HALD · Statistical Tables and Formulas
HALD · Statistical Theory with Engineering Applications
HARTIGAN · Clustering Algorithms
HILDERBRAND, LAING, and ROSENTHAL · Prediction Analysis of Cross Classifications
HOEL · Elementary Statistics, *Fourth Edition*
HOLLANDER and WOLFE · Nonparametric Statistical Methods
HUANG · Regression and Econometric Methods
JAGERS · Branching Processes with Biological Applications
JESSEN · Statistical Survey Techniques
JOHNSON and KOTZ · Distributions in Statistics
Discrete Distributions
Continuous Univariate Distributions-1
Continuous Univariate Distributions-2
Continuous Multivariate Distributions
JOHNSON and KOTZ · Urn Models and Their Application: An Approach to Modern Discrete Probability Theory
JOHNSON and LEONE · Statistics and Experimental Design in Engineering and the Physical Sciences, Volumes I and II, *Second Edition*
KEENEY and RAIFFA · Decisions with Multiple Objectives
LANCASTER · An Introduction to Medical Statistics
LEAMER · Specification Searches: Ad Hoc Inference with Nonexperimental Data
LEWIS · Stochastic Point Processes
McNEIL · Interactive Data Analysis
MANN, SCHAFER and SINGPURWALLA · Methods for Statistical Analysis of Reliability and Life Data
MEYER · Data Analysis for Scientists and Engineers
OTNES and ENOCHSON · Digital Times Series Analysis
PRENTER · Splines and Variational Methods
RAO and MITRA · Generalized Inverse of Matrices and Its Applications
SARD and WEINTRAUB · A Book of Splines
SEARLE · Linear Models
THOMAS · An Introduction to Applied Probability and Random Processes
WHITTLE · Optimization under Constraints
WILLIAMS · A Sampler on Sampling
WONNACOTT and WONNACOTT · Econometrics
WONNACOTT and WONNACOTT · Introductory Statistics, *Third Edition*
WONNACOTT and WONNACOTT · Introductory Statistics for Business and Economics, *Second Edition*
YOUDEN · Statistical Methods for Chemists
ZELLNER · An Introduction to Bayesian Inference in Econometrics

Tracts on Probability and Statistics

BHATTACHARYA and RAO · Normal Approximation and Asymptotic Expansions
BILLINGSLEY · Convergence of Probability Measures
JARDINE and SIBSON · Mathematical Taxonomy
RIORDAN · Combinatorial Identities